T0301573

The Changing Politics of Organic Food in
North America

The Changing Politics of Organic Food in North America

Lisa F. Clark

Department of Bioresource Policy, Business and Economics, University of Saskatchewan, Canada

Edward Elgar
PUBLISHING

Cheltenham, UK • Northampton, MA, USA

Published by
Edward Elgar Publishing Limited
The Lypiatts
15 Lansdown Road
Cheltenham
Glos GL50 2JA
UK

Edward Elgar Publishing, Inc.
William Pratt House
9 Dewey Court
Northampton
Massachusetts 01060
USA

A catalogue record for this book
is available from the British Library

Library of Congress Control Number: 2015935898

This book is available electronically in the Elgaronline
Social and Political Science subject collection
DOI 10.4337/9781784718282

ISBN 978 1 78471 827 5 (cased)
ISBN 978 1 78471 828 2 (eBook)

Typeset by Columns Design XML Ltd, Reading
Printed and bound in Great Britain by T.J. International Ltd, Padstow

Contents

Acronyms and abbreviations

3Q	third quarter
4Q	fourth quarter
AAFC	Agriculture and Agri-Food Canada
ADM	Archer Daniels Midland
AFL/CIO	American Federation of Labor/Congress of Industrial Organizations
AMS	Agriculture Marketing Service
BC	British Columbia
BEL	Belgium
CAD	Canadian dollars
CAQ	Conseil d'accréditation du Quebec
CBAN	Canadian Biotechnology Action Network
CBC	Canadian Broadcasting Corporation
CCOF	California Certified Organic Farmers
CDC	Centers for Disease Control and Prevention
CEC	Commission for Environmental Cooperation
CFIA	Canadian Food Inspection Agency
CGSB	Canadian General Standards Board
CITES	Convention on International Trade in Endangered Species of Wild Fauna and Flora
CNSOA	Canadian National Standards for Organic Agriculture
COAB	Canadian Organic Advisory Board
COABC	Certified Organic Association of British Columbia
Codex	Codex Alimentarius Commission
COG	Canadian Organic Growers
COPS	Canadian Organic Production Systems
COR	Canadian Organic Regime
COSA	Canadian Organic Soil Association
COUP	Canadian Organic Unity Project
CPB	Cartagena Protocol on Biosafety

CS	Celestial Seasonings
CSA	community-supported agriculture
CUSFTA	Canada–United States Free Trade Agreement
DDT	dichloro-diphenyl-trichloro-ethane
DEFRA	Department of Environment, Food and Rural Affairs
DSB	dispute settlement body
EC	European Community
EEC	European Economic Community
EPA	Environmental Protection Agency
ERS	Economic Research Service
EU	European Union
FAO	Food and Agriculture Organization
FBC	Fédération Biologique du Canada
FDA	Food and Drug Administration
FFCF	farm folk/city folk
FLO	Fair Trade Labelling Organization
FRA	France
FTC	Federal Trade Commission
GATT	General Agreement on Tariffs and Trade
GM	genetic modification
GMO	genetically modified organism
GOMA	global organic market access
HRSDC	Human Resouces and Skills Development Canada
IBS	International Basic Standards for Organic Production and Processing
ICFTU/ITS	International Confederation of Free Trade Unions/International Trade Secretariats
IFOAM	International Federation of Organic Agriculture Movements
IMF	International Monetary Fund
ISEAL	International Social and Environmental Accreditation and Labelling Alliance
ISO	International Organization of Standardization
MAB	Le Mouvement pour l'agriculture biologique
MEN	*Mother Earth News*
NAAEC	North American Agreement on Environmental Cooperation

NAALC	North American Agreement on Labor Cooperation
NAFTA	North American Free Trade Agreement
NASS	National Agriculture Statistics Service
NFU	National Farmer's Union
NGOs	non-governmental organizations
NL	the Netherlands
NOFA	Northeast Organic Farmers Association
NOP	National Organic Programme
NOS	National Organic Standards
NOSB	National Organic Standards Board
NPK	nitrogen, phosphorus and potassium
NSOA	National Standards for Organic Agriculture
NSMs	new social movements
NTB	non-tariff barrier
NYT	*New York Times*
OCA	Organic Consumers Association
OECD	Organisation for Economic Co-operation and Development
OFC	Organic Federation of Canada
OFPA	Organic Food Production Act
OFPANA	Organic Foods Production Association of North America
OG	*Organic Gardening Magazine*
OGF	*Organic Gardening and Farming Magazine*
OPR	organic production regime
ORC	Organic Regulatory Committee
OTA	Organic Trade Association
POP	persistent organic pollutant
PPMs	processes and production methods
SAFE	Sustainable Agriculture Food and Environment (Alliance)
SAWP	Seasonal Agricultural Workers Programme
SCC	Standards Council of Canada
SPS	sanitary and phytosanitary measures
TAN	transnational advocacy network
TBT	Agreement on Technical Barriers to Trade
TNC	transnational corporation
UK	United Kingdom
UNCTAD	United Nations Committee on Trade and Development

UNF	United Natural Foods
US	United States
USCB	United States Census Bureau
USD	United States dollars
USDA	United States Department of Agriculture
USDL	United States Department of Labor
USDS	United States Department of Statistics
WEC	*The Whole Earth Catalog*
WFM	whole foods market
WHO	World Health Organization
WTO	World Trade Organization

Acknowledgements

In writing this book I have encountered immense generosity among experts in the fields of politics, agriculture and food studies. The encouragement I have received over the years from colleagues, friends and family is by far the most important factor in getting this book published. I would like first to thank my senior doctoral supervisor, Marjorie Griffin Cohen as well as my committee members Stephen McBride and Theodore Cohn, who provided me with unwavering support and mentorship when I first embarked on the research for this book. I am grateful for Peter W.B. Phillips' advice and enthusiastic encouragement that kept me going even when I felt like just moving on. A special thank you goes to William A. Kerr who has been an invaluable resource throughout the process. I would like to particularly thank Daniel Beland, Kari Doerksen, Erika Dyck, Jill E. Hobbs, Julie Kaye, Cory Jansson, Stella Marinakis, Michele Mastroeni, Sara McPhee-Knowles, Jaime Leonard, Michael Plaxton and, last but not least, Camille D. Ryan for their thoughtful advice and insights. A huge thank you goes out to Natasha D. Patterson, who has always offered important guidance and never stopped believing that I could do this. Much love to my parents Patricia and Ralph Clark who have always supported me. Finally, I would like to thank Neil Hibbert. This book would literally not have been possible without his bottomless well of encouragement and patience, as well as his phenomenal editing skills. This book is what it is because of his amazing ability to help me see the 'forest through the trees' when all I could see were trees. To him, I am eternally grateful.

1. Introduction

In 2012, the global market for organic products was worth over $63.8 billion (USD), quadrupling its value from 1999 ($15.2 billion) (Willer and Lernoud, 2014:23). The United States (US) and to a lesser extent Canada have been the fastest growing markets for organic foods in the world. Between 2005 and 2010, the size of the market for organic products in both Canada and the US more than doubled (Haumann, 2014:242). In 2012, sales of organic food in the US were worth over $27 billion (USD) with over 40 million people purchasing some type of organic food in that year. According to a study quoted by the US Department of Agriculture (USDA), this figure has more than doubled from less than a decade ago (Osteen et al., 2012). To the north, Canada's comparatively smaller market topped $3.5 billion (CAD) in sales in 2012 (Holmes and Macey, 2014:249).

Organic food can be found everywhere from the local farmer's market to the aisles of Walmart across Canada and the US as consumer demand continues to rise. In the minds of some, the word organic conjures up images of clean, green and healthy food and is considered superior to conventional fare. Despite the 2008 global recession's reverberating effects on rates of unemployment and on food and fuel prices, organic food sales continue to grow as concerns over environmental toxins in the food system, 'frankenfoods' and worries about obesity, health and nutrition increase among the general population (Lernoud et al., 2014:251). To match growing demand, the land devoted to organic agriculture continues to rise across North America (Figure 1.1). Though land conversions slowed in the post-recession period, 2012 marks a noticeable increase from a decade ago.

Globally, the number of organic producers sat at 1.9 million in 2012 (Willer and Lernoud, 2014:23). Counted among them are an increasing number of Canadian and American producers. Based on census data from both countries, Figure 1.2 clearly shows that more individuals are participating in producing organic fruits, vegetables, grains, livestock and dairy products than ever before.

The current trend of market expansion and surging consumer demand is a far different reality from the one predicted for organic food in the

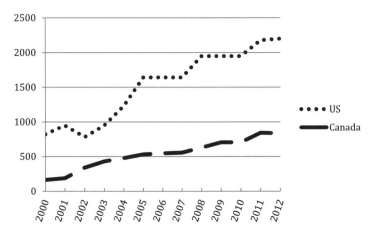

Sources: Green and Kremen, 2003; USDA, 2006; Macey, 2004; Willer and Yuseffi, 2004; Willer and Kilcher, 2009; Willer and Lernoud, 2014. US data for 2006, 2007, 2009, 2010 are based on Census data from 2005 and 2008. The data for 2010 is based on the previous year for Canada.

Figure 1.1   *Organic agricultural land in North America (in thousands of hectares) by year*

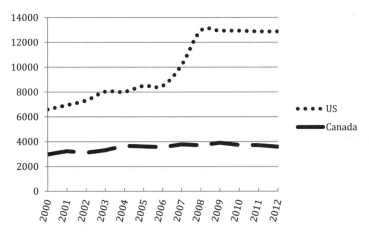

Sources: Willer and Yuseffi, 2004, 2005; Willer and Kilcher 2009; Willer and Lernoud, 2014. US figures for 2006 and 2009, 2010 and 2012 are estimates based on available Census data.

Figure 1.2   Total number of certified organic producers by year

1970s. Many government officials, scientists and media outlets believed that organic agriculture was inefficient, fraudulent and sometimes even dangerous to human health. Fears were voiced about what the expansion of organic agriculture could mean for the security of US food supply. As US Secretary of Agriculture Earl Butz argued in 1973, 'before we go back to organic agriculture in this country, somebody must decide which 50 million Americans we are going to let starve or go hungry' (quoted in Belasco, 1989:119). The American Association for Advancement of Science echoed Butz's concerns while debating the social implications of the 'organic food myth' at their 1974 annual meeting. Speakers discussed how 'food faddists, eccentrics and pseudoscientists were frightening the public into purchasing higher priced organic food by claiming organic foods were more nutritious, when there was no scientific evidence behind these claims'(*Washington Post*, 1974:F2). These fears and concerns have not materialized and instead, more and more Americans and Canadians are purchasing organic foods because organic foods are understood to embody qualities that food produced in a conventional manner does not.

Despite the remarkable success of the organic sector in recent decades it has been accompanied by increasing confusion over what certified organic actually means, and diverse values are associated with the idea of organic. For example, a survey conducted in 2010 of Canadian purchasers of organic foods revealed that while consumers understand the broader environmental issues related to organic agriculture, 34 percent of those surveyed believed organic food was better tasting and more nutritious than conventionally produced foods, while one-third of consumers believed organic agriculture uses no pesticides in production whatsoever (Campbell, Mhlanga and Lesschaeve, 2013:537).[1] A more recent study conducted in 2014 by US-based brand consultant BFG revealed that while 70 percent of those surveyed purchased organic foods, only 20 percent could define what the word 'organic' appearing on a food label actually meant. In explaining the results, the CEO of BFG said 'consumers are ultimately idealists… They want to believe. They trust the label and they're willing to pay more … for something … even though they are not totally sure what it means' (Brownstone, 2014).

The definitions of organic found in the standards and guidelines of governmental agencies, certifiers, industry associations and non-governmental organizations are much more complex and diverse than the

---

[1]   The regulations for organic production practices allow for the application of some pesticides and other preventative measures like antibiotics if an animal becomes ill, but there are strict requirements about what types can be used (see CAN/CGSB-32.331-2006; AMS-NOP-13-0011).

generally held public views expressed in consumer surveys. The guidelines of some certifiers can address the treatment and welfare of animals, the size and scale of farming establishments as well as labor standards covering agricultural workers. The confusion over the meaning of organic by those who buy organic foods but are not overly familiar with the intricacies of production guidelines stems from the longstanding ambiguity over a clear link between the principles and values commonly associated with organic agriculture and the actual practices that are involved in producing organic food according to formalized standards. What the word 'organic' actually refers to on a food label that is monitored and regulated by governmental agencies may be quite different from the values and principles consumers associate with the organic movement and organic food in general.

Though consumers today primarily purchase organic because it is perceived as a healthier, more nutritious and environmentally conscious option compared to conventional foods, consumer perspectives may not necessarily reflect the realities of organic production methods, or the nutritional qualities of organic foods. The ongoing debates about what defines organic agriculture by organic practitioners, businesses, government, consumers, academics and activists have their roots in an almost century-long evolution of the organic movement and the market for organic foods. What and who defines organic, and how various understandings influence the politics of organic food, have been important factors in shaping the development of the organic sector in North America. The current situation must be understood in relation to the series of reactions by the organic community to the interconnected social, economic and political changes in the broader food system. The result is a historically ambiguous position taken by the organic community towards defining a clear and coherent position on which social and economic relationships and processes are acceptable in organic agriculture practices and which are not. Today 'organic' is a highly contested concept such that its very meaning continues to spark social and political debates among those who are concerned about food quality, safety, nutrition and ecological sustainability.

THE ROOTS OF ORGANIC

From its early beginnings, organic agriculture has defined itself in relation to the structures and practices of industrial agriculture (often referred to as 'conventional' agriculture). In many ways the principles associated with organic agriculture emerged in response to the structures and practices of the conventional food system. Despite the many benefits attributed to industrialized food production, including a more stable food supply and lower food prices, concern over the environmental implications of the use of synthetic chemicals for food production, including depleted soil fertility, nutrient run-off and water system pollution from farm waste, were motivating factors for early organic practitioners to experiment with agricultural techniques (Rodale, 2010). The desire to maintain soil health in food production remains the central principle of the organic approach, finding its way into formal and informal organic production standards and regulations around the world. Organic food is often associated with a more environmentally friendly form of food production because organic agriculture rejects many fossil fuel intensive agricultural technologies used in conventional production that are proven to cause environmental damage. This association has contributed to growth in markets for organic foods and has also been an important link of principles between the organic and environmental social movements in the 1960s.

Industrializing processes also had tremendous impacts on the socio-economic organization of agricultural production in North America and informed the development of the organic sector. With the increasing technological intensity of production methods and rising demand for higher farm outputs at lower costs, the post-war period saw a major shift in how the agricultural sector was organized. Many of the family farms in North America faced difficulties maintaining competitiveness as international commodities prices fluctuated and pressures to reduce costs increased (Bonanno et al., 1994; Kloppenburg (ed.), 1988; Goodman, Sorj and Wilkinson, 1987; Dahlberg, 1979). Some early organic practitioners viewed the social and economic changes resulting from increased use of industrial agricultural technologies as evidence of the growing control of agribusiness in the food system and the expansion of unsustainable farming practices. The socio-economic critiques of industrialized food production, and the desire to remain independent from reliance on agribusiness, motivated many practitioners to experiment with on-farm waste recycling techniques and non-chemical pest management strategies. These practices served as an ideational foundation for early definitions of organic agriculture and were important components to the growth and expansion of the organic movement and organic agriculture across North America throughout the twentieth century (Peters, 1979; MacRae, 1990; Reed, 2010).

Consumer awareness of the human and environmental health and safety hazards associated with industrial agriculture, that plays such an important role in today's market expansion for organic foods, has its roots in the new social movements of the 1960s. Publicized instances of compromises to food safety and environmental damage attributed to the over-use of synthetic pesticides and industrial processes (such as Rachel Carson's *Silent Spring* (1962)) caused growing concern about the health and safety implications of conventional agriculture. This continued with well-publicized food contaminations and safety hazards like the Alar apple controversy in the 1980s, the BSE crisis in Europe in the 1990s, the foot-and-mouth disease outbreak in the UK in the 2000s, as well as food recalls ranging from ground beef to bagged spinach in the US (Lang, 2004; Nestle, 2004). Increasing public awareness of health and safety issues related to food in Canada and the US helped to put pressure on both government and businesses involved in the agri-food system to be more accountable and proactive when hazards were identified, resulting in the creation and expansion of food safety inspections, testing and monitoring by government agencies. Despite ongoing efforts to reduce food safety risks throughout the food system, challenges continue. In 2007, the World Health Organization (WHO) estimated that 30 percent of the population in industrialized countries suffers from food-borne illnesses every year, while 76 million cases of food-borne illness were reported annually in the US (WHO, 2007). More recently, the US Center for Disease Control and Prevention (CDC) estimates that there are 48 million cases, 128 000 hospitalizations and 3000 deaths related to food-borne illnesses in the US every year (CDC, 2013). Consumer fears surrounding food safety have helped expand the market for organic foods that are promoted as safer and healthier than conventionally produced food (Food and Agriculture Organization (FAO), 2003; Macey, 2007).

Prior to the 1990s, networks of small-scale organic producers and businesses served a relatively small consumer base in Canada and the US. In the 1970s, there were fewer than 4000 health food stores, the primary sales point for organics at the time, in both Canada and the US (Myers, 1976; Cooper, 2006). But concerns over the environmental and human costs of industrialized agriculture, the use of biotechnology in the food system (dubbed as 'unnatural', tools of agribusiness and threats to biodiversity), and food safety fears fueled the explosive consumer demand for organic products across North America in the 1990s.

The rapid rise in demand triggered significant changes in the structure of the organic sector. The two most notable and impactful are the involvement of what were traditionally viewed as 'conventional corporate interests' and the creation of formalized regulations for 'certified' organic production processes. Conventional corporate interests met rapid consumer demand with a flurry of investment, resulting in many corporate mergers, acquisitions and the introduction of new brands of organic food products. Market expansion also necessitated the need for more formalized regulations and standards to monitor and evaluate organic agricultural processes and foods transported across North America, and thus regulatory frameworks were developed in both the US and Canada throughout the 1990s and into the 2000s. The rules governing activities in the organic sector became part of the broader regulatory frameworks for food and agriculture originally designed for conventional means of production. This further distanced some of the more substantive elements of the organic principles from organic practices because of the overarching principles and mandates informing agriculture and food policy (increasing crop yields, international market expansion, trade-related product equivalence regulations, etc.).

The rush to capitalize on growing consumer demand through investment by conventional agri-food corporations like Coca-Cola and General Mills caused heated debates among members of the organic community regarding how this happened and what it meant for the future of organic food as an alternative to conventional foods. In the early 2000s, Julie Guthman (2004:10) observed that:

> the organic food sector is increasingly bifurcated into two very different systems of provision: one producing lower cost and/or processed organic food … the other producing higher value produce in direct markets and appealing to meanings of organicism, political change and novelty … Practitioners in both systems are able to claim the moral high ground.

The lack of a firm commitment to a particular political platform with consistent goals of socio-economic change (or at least an alternative to market relations in the conventional food system) has meant that the meaning of organic has been continually reinterpreted in relation to broader changes in the conventional food system. It has also meant that diverse approaches can co-exist within the organic sector that do not necessarily adhere to any social or political principles historically associated with the organic movement.

The implications of formalizing into public policy the way organic food is produced, regulated and monitored for the integrity of organic principles and the organic movement have been explored from myriad academic perspectives. Some discussions focus on the evaluation of innovations in organic agriculture in terms of organization and economic productivity, detailing techniques, methods and implications for outputs (Cacek and Langner, 1986; Lampkin and Padel (eds), 1994; Bellon and Penvern (eds), 2014). Country and farm level analyses have provided important insights into the changes experienced by producers and consumers as increased demand puts pressure on maximizing outputs. There is also ongoing interest in the dynamics of the regulatory environment for organic foods, inspection and certification systems that explore how organic standards are developed, how they relate to other forms of food regulation and how they are evolving, including an entire volume of the journal *Food Policy* (42) in 2013 (Riddle and Coody, 2003; Bostrom and Klintman, 2006; Mutersbaugh, 2005; Padel, 2009; Daugbjerg, 2012). Others focus on the dynamics of consumption, consumer-led food movements and the socio-political implications of expanding organic markets (Lohr, 1998; Allen and Kovach, 2000; Murdoch and Miele, 1999; Klonsky, 2000; Buck et al., 1997; Raynolds, 2004; Howard, 2009b; Johnson and Szabo, 2011). How consumption habits and organic agricultural production practices influence the shape of the organic sector and whether these changes have moved the organic model towards resembling those found in the industrialized food system has occupied a significant space in academic discussions (Clunies-Ross, 1990; Guthman, 2004; Hall and Mogyorody, 2001; DeLind, 2000; Campbell and Rosin, 2011).

The dichotomized 'conventionalization versus differentiation' debate has begun to soften somewhat to account for the diversity of styles and approaches to organic production. These frequently deviate from a purely principled understanding of organic on the one hand, to a corporatized, factory-farm model of agricultural practices on the other. Instead, current discussions increasingly focus on the nuances of organic production, consumption, activism and regulation in relation to the social and political dynamics of the broader food system to demonstrate how organic continues to differentiate itself from conventional production as well as diversities within organic sectors (Conford, 2001; Reed, 2010). Though in some academic circles the debate over whether organic has 'sold out' to corporate interests in North America is old news, other discussions continue to chronicle and attempt to understand the evolving regulatory system and changing consumer demands for organic foods in attempts to preserve a degree of 'organic integrity' (Hoodes et al., 2010). The perception that all organic food is, by definition, more nutritious and more ethically sound because it does not contain genetically modified organisms (GMOs) and chemical-inputs continues to flourish in North America. For this reason, the relationship between the principles and practices in the organic food sector, and their relationship to the conventional food system, remains timely and relevant.

THE POLITICS OF ORGANIC: INSTITUTIONALIZATION AND CHANGE

This book examines how the evolving organic movement cultivated a set of principles that have been continuously reinterpreted and redefined by practitioners, businesses, policy-makers, advocates and consumers in North America. It focuses on the relationship between the organic movements in Canada and the US, the economic connections between organic markets on both sides of the border and how both countries' organic food regulations relate to one another. In many contemporary studies of organic food it is often the American and European cases that are used for comparison because of similarities in the size of their markets for organic products and their mutual importance as major trading partners in food products (Bostrom and Klintman, 2006; Klintman and Bostrom, 2013). The politics of organic food in Canada has received much less scholarly attention despite the presence of the early organic movement in the 1950s and the involvement of US agri-food corporations in the Canadian food sector. Canada is also a major exporter of organic food products to both the US and Europe, while it imports a significant proportion of organic food sold domestically from the US. In 2008, it was estimated that Canada imported 74 percent of all retail organic products from the US (ACNeilson Canada, 2009), indicating a close and significant relationship between each country's organic sectors.

Unlike many other scholarly examinations of organic agriculture, this book does not debate the nutrition and health claims attributed to organic foods or the agronomic aspects of organic agriculture. Instead, it focuses on the politics of organic agriculture and how the organic movement emerged to challenge the socio-economic relationships found in conventional agriculture over time. It examines how the principles of organic and the increasing presence of organic food in mainstream markets in the 1970s and 1980s paved the way for the dramatic market expansion and conventional corporate involvement in the business of organic foods in the 1990s. The book explores how the development of an alternative food movement that prided itself on its diverse membership and independence from government and agribusiness was influenced by many practices and behaviors found in industrial agricultural sectors. The case of how the organic movement changed in response to growing mainstream popularity of organic food can shed some light on how alternative social movements change and evolve in relation to broader societal changes and how processes of institutionalization create pressures for altering the movement's goals (see Fridell, 2007). *The Changing Politics of Organic Food in*

North America shows that attempts to achieve socio-economic change through conventional market mechanisms increases a social movement's susceptibility to the involvement of outside actors who want to profit from the association of a purchasable good with the values and principles consumers associate with the movement. It also provides a better understanding of current and emergent food movements and how they can attempt to meet their goals within the current globalized food system.

This book explores the changing politics of organic food by examining the processes of institutionalization to uncover why certain principles and practices associated with organic were included in conventional policy frameworks, corporate business strategies and the objectives of social activism, and why others were not. It discusses how vocal members (both individual and organizations) associated with the organic movement responded to the growing markets for organic foods and the institutionalization of organic agriculture into pre-existing regulatory frameworks covering agricultural practices and food labeling and safety. Institutionalization is a process of formalizing principles, practices and normative behaviors as public rules, standards and decision-making structures to improve the predictability of behaviors among stakeholders within a particular system (North, 1990). Institutionalization is a way to explain how and why rules, structures and institutions themselves are created, maintained and modified by evaluating relationships and processes within a given system. It is a useful way to explore the politics of organic food from a twenty-first century perspective because it not only considers how and why the behaviors of stakeholders and institutions change, but it also attends to changing relationships between the constellation of stakeholders, institutions and the ideas they promote to help explain outcomes. It also provides a way of understanding why some substantive principles guiding production that address the negative social, political and ecological outcomes of industrialized agriculture were not included in standards and formalized definitions of organic, and why other practices associated with industrialized agriculture, like synthetic inputs and GMOs, became central focal points of organic regulations in Canada and the US.

Over eight chapters, *The Changing Politics of Organic Food in North America* explores how the involvement of advocates, practitioners, consumers, governments and corporations in building formal definitions of organic have influenced the evolving politics of organic food in Canada and the US. Chapter 2 begins by identifying the transformative effects industrialization had on agriculture in Canada and the US throughout the twentieth century. It then explores two contending conceptions of organic – described as the 'process-based' and the 'product-based'. The process-based definition is built upon a combination of interconnected social, ecological and economic principles that are meant to guide practices, and which were evident in the early organic movement. This definition is traditionally associated with organic agriculture. The competing 'product-based' definition is more reflective (rather than prescriptive) of the practices found in organic production systems, though it shares some characteristics with the process-based definition. It has become the dominant approach in practice because of limited attention given to outlining and formalizing preferred types of market relations and economic organization in the process-based definition of organic (Clark, 2007). This conception is more instrumental in nature by virtue of emphasizing the practices that directly influence the material qualities of the end product, such as restrictions of synthetic inputs like GMOs. The chapter demonstrates that as a result of a lack of clarity and emphasis on crucial factors in the development of organic principles, the process-based definition of organic failed to make serious headway in the 1970s as the market for organic products expanded.

Chapter 3 traces the historical development of corporate activity in the conventional and organic agri-food sectors in Canada and the US in relation to the development of the product-based definition of organic. Corporate actors have applied many competitive business models found in the conventional agri-food sector to the organic sector. The discussion identifies three dominant corporate strategies applied to the organic agri-food sector starting in the 1980s: consolidation of ownership through mergers and acquisitions, strategic alliances and brand introduction. The successful application of these strategies has influenced patterns of ownership within the organic sector. It has also influenced how organic food is regulated and monitored in the food system.

The development of organic standards and regulations at multiple levels of governance has significantly changed various aspects of organic agriculture since the 1980s, effectively entrenching the product-based conception. Chapter 4 looks at how organic standards originally developed at the local level through producer-based associations became part of regulatory frameworks designed for conventional agriculture. It traces the creation of national standards, examining how various actors influenced American and Canadian regulatory frameworks. It also explores how regulations were designed to incorporate organic agriculture into the global trade regime by diminishing the role of process in the evaluation of organic foods, which limits the ability to include socio-economic principles in formalized definitions. As the market for organic products expanded, top-down forces of regulation and harmonization of standards in the global trade regime imposed downward pressures on bottom-up efforts to include more process-based features to the evaluation of organic foods. Chapter 5 expands the discussion surrounding policy to include international trade agreements such as the North American Free Trade Agreement (NAFTA) and those administered through the WTO (i.e., the Agreement on Technical Barriers to Trade (TBT) and the Sanitary and Phytosanitary Measures (SPS) Agreement), to explore how these regulations apply to organic food circulating in the global economy. It takes a closer look at the role 'product equivalence' has in the rules and standards applicable to traded organic foods and how the concept of 'like' products in trade agreements influences how organic foods are evaluated in relation to formal standards.

Chapters 6 and 7 explore the changing political dynamics of the organic movement from its beginnings until the 2010s. Chapter 6 uses the policy process model to examine the actors, institutions and ideas that composed the early organic movement (McAdam et al., 1996). It argues that a number of other social movements emerging in the 1960s, such as the environmental and sustainable agriculture movements, helped the organic movement to expand its membership and broaden its appeal to the mainstream. It also discusses how, as professional organizations entered the organic sector (that coincided with the rise of the product-based approach), its status as a social movement changed. Chapter 7 addresses the dramatic influence that corporate actors had on the objectives and structure of the organic social movement, presenting the evolution of the contemporary organic movement and how it has moved away from being classified as a textbook 'social movement' as the product-based definition of organic gained ground. The current incarnation of the movement is better understood from an advocacy network perspective (Keck and Sikkink, 1998; 2000) that involves a variety of actors (governments, corporations) that are typically the target of social protest. The inclusion of these new actors had significant implications for the applicability of the process-based definition of organic in the mainstream in the 1980s, yet there continues to be a social movement contingent within the broader network that remains committed to the process-based definition of organic. Chapter 8 concludes by taking a closer look at the current status of the organic movement and the organic market as organic food has taken its place in the mainstream food system. It discusses what lessons can be drawn from the evolving relationships between actors, institutions and ideas associated with organic food and suggests some possible futures for the politics of organic food and other food movements as demand continues to expand for alternatives to conventional food in a globalized food system.

2. A clash of values: competing definitions of organic

INTRODUCTION

Organic agriculture is understood to be a form of food production that differentiates itself from conventional agriculture on the basis of a set of principles. Though attention to soil health was the original impetus for the development of organic agricultural methods and techniques, organic agricultural practices quickly developed a broader set of principles designed to instruct practices. The principles guiding organic food production were originally defined in opposition to the rise of conventional industrial practices premised on economic efficiency and increasing yields. The way organic agriculture has been incorporated into the mainstream agri-food system, however, presents challenges to the integrity of many of the fundamental principles and values associated with the organic food movement. What was once a relatively unified challenge to industrial methods of food production is now made up of two contending, yet not consistently oppositional interpretations of the definition of organic; the *process-based* and *product-based* definitions.

There are many specific approaches to organic agriculture that do not fit neatly into either definition discussed here, but the two overarching definitions of organic generally reflect the different programmatic commitments and practices we see in Canada and the US today. Others have noted two distinct approaches to organic production existing in the North American context and elsewhere. Klintman and Bostrom (2013:108) label the two approaches as 'behind the shelf' and 'on the shelf'. The first refers to a method of production that accounts for social and cultural processes not necessarily tangible in the final, end product. The second refers to a form of production focused on the qualities of the end product at the consumption stage. While Klintman and Bostrom's analytical focus is on the product, Hall (2014:401) primarily distinguishes between 'deep' and 'shallow' organics in terms of whether organic production systems take a holistic, inclusive approach that pay attention to psycho-social and cultural aspects related to food production, or whether the production system uses input substitution to fulfill the formal certification requirements. While both of these categorizations are valuable to distinguishing between types of organic food within organic systems, this chapter takes a broader perspective of the process-based and product-based definitions. It looks at how definitions developed in terms of their relationship to the practices and normative behaviors in the conventional food system. Although dichotomizing current organic practices and ideologies is problematic since a large 'gray area' exists, it is helpful to outline the several major differences between an approach to organic agriculture attempting to internalize social and environmental costs throughout the production process, and the other approach defining organic by product qualities that pay less attention to social and political aspects of food production processes. This chapter begins by presenting the historical trajectory of the process-based definition and its specific points of emphasis on the value created at various locations in the production process. It then discusses the product-based definition that emerged as markets for organic food expanded beyond the farm gate. The product-based definition places far less emphasis on substantive social, economic and ecological goals throughout the production process.

THE PROCESS-BASED DEFINITION OF ORGANIC

The first time a published work used the word 'organic' to describe agricultural processes was in 1940. *Look to the Land*, written by Briton Lord Northbourne, conceptualized the farm as a type of organism, where each part worked together as part of a greater system, something that contrasted with emergent forms of industrialized agricultural production (Kirschenmann, 2004). The idea that the way food was produced was as important as the food itself served as the conceptual basis for a set of organic principles that viewed food as distinct from other commodities. The process-based definition of organic that would be a fundamental component to the more traditional way of practicing organic agriculture emerged from decades-long experimentation and development of a closed food production system. It would characterize the early organic movement that consisted of a small group of individuals committed to developing a sustainable way to produce food by focusing on maintaining the health of the soil. Supporters of this approach mounted a reaction to structural and processional changes in the agri-food system considered to be environmentally and socially unsustainable. Those considered 'radicals' at the time criticized industrial agriculture for its reliance on fossil fuels and the implications for soil health, as well as the social and environmental costs associated with industrializing processes. Early advocates of organic agriculture sought to internalize and reduce costs and reliance on off-farm inputs in an attempt to develop a sustainable agri-food system. By attempting to internalize the production of soil inputs and the social and environmental process costs in the overall price of the end product, principled organic agriculture sets itself apart from conventional forms of agriculture by stressing the importance of process to various qualities of the end product. Nicholas Lampkin (1994:4), a pioneering scholar studying the economic dimensions of organic agriculture, conceptualizes organic agriculture along process-based lines:

> [organic is] an approach to agriculture where the aim is: to create integrated, humane, environmentally and economically sustainable agricultural production systems, which maximize reliance on farm-derived renewable sources and the management of ecological and biological processes and interactions, so as to provide acceptable levels of crop, livestock and human nutrition, protection from pests and diseases, and an appropriate return to the human and other resources employed.

This academic interpretation of what the process-based definition of organic covers is prescriptive, yet it also defines a set of goals and values associated with organic agriculture and the organic movement. It does not explicitly define the elements of organic agriculture in opposition to the elements of industrial agriculture, but rather highlights many of the key aspects of organic agriculture independently from the industrial model. A definition of organic devised by the International Federation for Organic Agriculture Movements (IFOAM) emphasizes aspects similar to Lampkin, but in a more methodical way. The formal definition IFOAM (2009) promotes is as follows:

> [o]rganic agriculture is a production system that sustains the health of soils, ecosystems and people. It relies on ecological processes, biodiversity and cycles adapted to local conditions, rather than the use of inputs with adverse effects. Organic agriculture combines tradition, innovation and science to benefit the shared environment and promote fair relationships and a good quality of life for all involved.

IFOAM builds this definition of organic agriculture on four basic principles: health, ecology, fairness and care (Sligh and Cierpka, 2007:36). The principle of health applies to humans, animals and the environment, while the principle of ecology embodies ideas of supporting biodiversity and sustainability, and reducing environmental damage as a result of food production. The principle of fairness speaks to the social elements ascribed to organic agriculture. It defines organic agriculture as a type of food production that strives to achieve equality, respect and justice between people and the environment. Finally, IFOAM highlights the principle of care. IFOAM defines organic agriculture as a type of food production that should be practiced with mindfulness of future generations of people and the future of the environment (IFOAM, 2008). These principles provide guidelines for how organic agriculture should be practiced. Though the principles do not explicitly mention the technical requirements for a product to be considered organic, they do promote a set of values that establish the parameters for the process-based definition of organic.

IFOAM's principles, like Lampkin's definitional points, are instructive and normative, meaning they outline the goals that *should* be considered when practicing organic agriculture and not necessarily how it *is* practiced. Both definitions focus on the substantive qualities associated with organic production that is largely an issue of process. From these two related definitions of organic, three central principles can be identified: economic viability, social sustainability and environmental sustainability. Together, these interconnected principles make up the process-based definition of organic. Not all of these principles became part of the process-based definition at the same point in time, nor is it implied that all organic practitioners who value the production process as integral to the value associated with organic products subscribe to each tenet in the same way. There is far too much diversity in agricultural practices to attempt to capture all variations on organic techniques and methods in one definition. But to understand how these principles together create a broadly coherent definition of organic that in many ways continues to be associated with the practices of organic agriculture and the organic movement, the relationship between the economic, environmental and social dimensions of agricultural practices must be further explored. The environmental and economic concerns were the first to become significant characteristics of the process-based definition of organic, and the social elements became more overtly associated with organic food in the late 1960s. Despite the uneven development of the foundational principles, they are deeply interdependent and attribute normative value to the processes of organic agriculture.

The origins of the process-based definition that guide some forms of organic agriculture can be traced to pre-World War II Europe. As principles of organic agriculture began to formulate in Europe, early practitioners of organic methods in Canada and the US made up a small and concentrated movement in the 1950s (Conford, 2001). The values and goals of organic agriculture emerged as a form of resistance to the rapid changes in agriculture brought about by the institutionalization and the perceived negative environmental outcomes of industrialized agriculture. But before industrialization transformed agriculture, the majority of people around the world were practicing organic or natural agriculture simply because it was the only option available to them (Mitchell, 1975; McMichael, 2004). Although mechanization and chemicalization in agriculture occurred before the twentieth century, only after World War II did chemical farming become widely practiced by food producers in industrialized countries (Atkins and Bowler, 2001).

Industrialized agriculture can be conceived of as harnessing two innovative technologies of the early twentieth century that, when used together, increase farm yields and production efficiency through standardizing processes (Millstone and Lang, 2003:56). Farmers have struggled with securing crop yields and the quality of the harvest for centuries. A balance of the macronutrients nitrogen, phosphorus and potassium (NPK) must be present in the soil in order for plants to grow properly. Traditionally, farmers relied on various sources to return NPK to the soil in the post-harvest period. Farmers have historically relied upon compost, manure or mined inorganic sources to replenish the soil. In non-chemical settings, a mix of NPK macronutrients are returned to the soil through the use of cover crops like alfalfa to provide nitrogen and the application of animal manure or compost to provide the phosphorus and potassium (Adamchak, 2008:16).

In 1840, a German chemist named Justus Leibig published a report entitled *Chemistry in its Application to Agriculture and Physiology* that discussed how fossil fuels could be processed to create the ideal balance of NPK to fertilize the soil (Conford, 2001:38). It is estimated that it requires the equivalent of 30 gallons of fossil fuel to produce enough synthetic NPK to grow an acre of corn (Shapouri et al., 2004). Synthetic fertilizers could then be applied to the field with controlled balances of NPK, which would assure farmers that the proper balance of these macronutrients were present in the soil. Synthetic fertilizers created a reliable and more convenient source of NPK for farmers, which helped to reduce some of the economic uncertainties of farming.

Using synthetic chemicals and practicing monoculture defined the paradigm shift towards industrialized agriculture. Chemical weapons used during World War II were found to be effective as defoliants, and they were also discovered to be very successful at killing pests and weeds. Dichloro-diphenyl-trichloro-ethane (DDT) was first developed in 1874, but it was not until 1939 that its effectiveness as an insecticide was discovered (Carson, 1962:20). Scientists demonstrated that by using chemicals like DDT to kill insects, higher agricultural crop yields could be attained (Clarke et al., 2001; McMichael, 2004).[1] The increasing use of fossil fuels to make a variety of products (including fertilizers and pesticides) and the usage of energy-intensive farming equipment were very effective at increasing farm yields while reducing the need for manual and animal farm labor (Steffen, 1972:5). These advances were welcomed as they lessened some of the economic risks associated with farming and bolstered efforts to achieve a reliable food supply. Although the scientific data regarding chemical fertilizers existed well before the end of World War II, the supply of crude oil and chemical warfare technology made chemical agriculture a viable economic option for farmers seeking to reduce the need for manual labor, to increase crop yields and to lessen some of the economic risks of farming. This new mode of food production appeared to be a panacea for hunger and the unpredictable forces of nature that could destroy crops and livelihoods. Industrialization, in short, was the way to achieve food security by enhancing control over inputs so as to have more predictable outputs.

Scientists and state agricultural departments promoted chemical agriculture as progressive, efficient and necessary to meet the food requirements of growing post-war populations. In an effort to assure national food security, governments invested heavily in their agricultural sectors by creating institutions and policies to stabilize and increase the quantities of agricultural foodstuffs. Previously barren soils were replenished with chemical fertilizers, which increased the acreage designated for food production. Farmers in Canada and the US continually increased their use of chemical fertilizers to boost the nutrients in soils, which increased yields. Figure 2.1 shows the increase in fertilizer usage in Canada and the US between 1961–71.

In addition to using chemical fertilizers, pesticides and herbicides, mono-cropping was essential to standardize outputs and increase productive efficiency. Monoculture refers to the planting of one type of crop in a designated area, and became the norm on farms across Canada and the US. Monoculture was understood by agricultural departments as the most efficient way of producing a high volume of crops in the shortest period of time (Berry, 1977; Skogstad, 1987; Kneen, 1989). Machinery was specifically designed for preparing the soil for harvesting a particular crop, making sowing and harvesting quicker and easier for farmers. Monoculture reduces the amount of time and labor necessary to grow and

---

[1]   DDT is labeled a persistent organic pollutant (POP) by the UN, and in 2001 the WHO/FAO drafted an agreement to limit its usage due to its proven damaging effects on human and environmental health (Clarke et al., 2001:13).

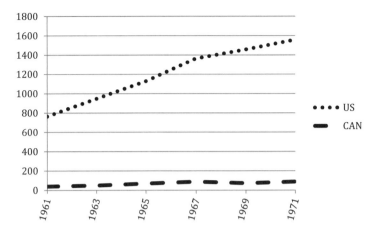

Source: FAO AGROSTAT database.

*Figure 2.1   Total fertilizer consumption in Canada and the US
            (10 000 metric tonnes) by year*

harvest successfully (Friedmann, 2000:492). Before monoculture most farms were polycultural and consisted of a mix of crops sown in the same area that were rotated from year to year. The logic behind crop rotation is that planting cover crops on a yearly rotational basis will replenish nitrogen to the soil (green manure), improving the soil conditions for next year's crops. By practicing crop rotation, the soil maintains its nutrient balance because the same nutrients are not leached from the soil continuously by the same type of crops. Though the way crop rotation is practiced radically changed as polyculture gave way to monoculture, largely because of mechanization, farmers continue to rotate crops today to aid in soil fertility and management.

Polyculture requires much more manual labor to maintain a weed and pest-free environment, and it is often practiced on small plots of land geared towards household consumption or regional markets. As mono-culture became the standard, sales of mechanical farm implements rose correspondingly. From 1969–73, farm implement sales in Canada grew from $410 million to $656 million (CAD), more than a 62 percent increase over a four-year period (Mitchell, 1975:62). Agricultural tech-nologies could free farmers from the physical labor required to produce crops, while increasing the profitability of their farms. The utilization of agricultural technologies on a wide scale is what truly defined the emerging industrializing processes of agriculture; technology effectively rearranged production processes to maximize efficiency, increased farm yields and transformed what it meant to be a farmer (Friedmann, 1991).

Though industrialized agriculture in many ways delivered on its promises of reducing risk and uncertainty regarding crop yields and increasing the profitability of farming, early organic practices attempting to return to a less mechanized way of producing food began to surface in Europe in the 1940s (Tate, 1994:11; Raynolds, 2004:735). Although countless individuals in both Europe and North America contributed to the basic normative principles of the process-based definition of organic (Peters, 1979), there were three notable contributors in the early twentieth century who were critical of industrialized agriculture and presented a coherent and direct challenge to what had quickly become 'conventional' agriculture. Britons Lady Eve Balfour and Sir Albert Howard, and American J.I. Rodale, were key early organic practitioners who championed the techniques and gave a voice to organic agriculture as a viable alternative way of producing food. They focused on the impact of chemical agriculture and the pursuit of economic efficiency on the health of the soil. All three organic practitioners would later help establish formal organizations and associations promoting organic agriculture, and their ideas would be referenced and put into practice by their followers.

Balfour was one of the first scientists to experiment with organic techniques as described in her work *The Living Soil* (1943). Balfour's research focused on how to replenish nutrient levels in the soil. She earned an Agricultural Diploma from Reading University in 1917 and became a champion of the organic techniques she found to be extremely beneficial to the health of the soil and plants (Conford, 2001:88). Balfour fundamentally challenged the notions put forth by other scientists at the time concerning the beneficial use of industrial chemical inputs and the argument that farming without synthetic fertilizers and pesticides produced low yields and were financially costly to the producer. Though it is commonly assumed that organic agriculture is strictly non-chemical, certain substances are allowed in preventing pest destruction, such as copper sulfate – the Bordeaux mix, a combination of copper sulfate, lime and water – typically used as part of an integrated pest management program. This mixture works to combat unwanted fungus and bacterial growth on crops, most commonly fruit and nut trees, and vine fruits. Since copper sulfate and lime are naturally occurring, they are not considered 'synthetic' among the organic agriculture community (Broome, 2012). Today, chemicals such as copper sulfate are industrially produced but do not qualify as synthetic chemicals because they are not petroleum-based, are naturally occurring and have been used for decades as fungicides and bactericides by organic practitioners (Ryan, 2001:16).

On Balfour's own experimental farm in England, she demonstrated that organic agriculture could be financially viable and could, in fact, save farmers money by recycling vegetable and animal waste to produce nutrient-rich humus to fertilize the soil (Balfour, 1943: Chapter 3). Her ideas about organic agriculture as a sound environmental *and* economic alternative to industrial agriculture questioned the role agriculture chemical companies played in food production, and viewed farmers' reliance upon them as financially costly and environmentally harmful to the health of the soil. Two years after *The Living Soil* was published, Lady Balfour, Sir Albert Howard and a number of other supporters of organic agriculture formed The Soil Association in Britain, which became a venue for discussing soil health and how to educate the public about its importance.

Sir Albert Howard is one of the most widely acknowledged pioneers of organic agriculture, and his work in promoting and practicing organic techniques helped to raise organic agriculture's profile as a viable alternative to industrial agriculture around the world (Merrill, 1976; Belasco, 1989). Howard was one of the first to make the causal connection between the health of the soil and the quality of food and human health by running a number of experiments with composting and crop rotations. *An Agricultural Testament* (1943) was one of the first books published that explored organic farming techniques. As a trained chemist, Howard was critical of what he called 'laboratory hermits' and their dislocation from what was happening 'in the field' (Berry, 1977:46). Howard decided that the best way to bridge the gap between the lab and the field was for him to experiment with agricultural techniques that worked with nature instead of trying to conquer it. Through his experiments, he found that he could successfully produce healthy crops and soil without synthetic chemical inputs.

Howard spent almost a decade in India (1924–31), experimenting with ways to add nutrients back into the soil without synthetic chemicals. In *An Agricultural Testament*, he introduced the Indore Process (named after the Indian state where the technique was developed), which is a composting method that uses vegetable waste and manure to produce nutrient-rich humus to fertilize the soil and crops (Howard, 1943: Chapter 4; Conford, 2001:246). With the publication of *An Agricultural Testament*, the Indore Process became a widespread method used to grow a number of agricultural crops such as rice, sugar cane, coffee and fruit. As a result of Howard's experiments and the connections he made between the health of the soil, crops and people, he cautioned that chemical-based agriculture had negative consequences for the environment and public health because it relied on external, non-renewable, chemical inputs for financial success (Vos, 2000:246).

In 1946, Howard published *War in the Soil*, which described the ongoing conflict between the idea that the agriculture should focus on producing fresh and natural foods and the desire for agribusiness to maximize profits. Though Howard's primary focus was on soil science in his published works, like his contemporaries, he felt the need to comment on the politics of the rapidly industrializing agri-food system at the end of World War II. *The Soil and Health: A Study of Organic Agriculture* (1947), another of Howard's publications, took on the negative outcomes produced by chemical agriculture. It denounced the exploitation of the environment for profit and stressed the importance of people having a role in how their food is produced. Howard promoted a system of 'farming which puts a stop to the exploitation of land for the purpose of profit', claiming that 'the electorate alone has the power of enforcing this and to do so it must first realize the full implications of the problem' (Howard, 1947:13). Howard became the first individual to associate with organic agriculture, arguing that industrialized, chemical agriculture focuses exclusively on the economic elements of agricultural production, basing decisions on the forces of supply and demand.

While he was developing the Indore Process in India, Howard made attempts at fair labor practices by providing laborers on his experimental farm with rest breaks, medical services, prompt payment and standard work hours. By using organic techniques and valuing the human labor inputs, Howard attempted to attach social benefits to his agricultural experiments at the Agricultural Research Institute in Pusa, India (Conford, 2001:55). Though it could be argued that the social benefits associated with the process-based definition of organic are purely products of the 1960s countercultural social movements, Howard's attempts at treating workers fairly even on a micro-scale had far-reaching influence. Organic practitioners routinely looked to Howard's texts for guidance, and his attempts at fair treatment of labor and criticisms of industrial agriculture did not go unnoticed. Not all organic practitioners would attempt or want to institute his practices toward workers or challenge conventional labor practices in agriculture, but Howard's documentation of the conditions of agricultural labor on his experimental farms points to a degree of social consciousness early on in the organic movement's history. Howard's attention to agricultural labor practices in his works helped to draw readers' attention to the social conditions related to farming.

One of the main arguments Howard put forward in his writings was the critique of industrialized modes of food production and its treatment of food as if it were the same as any other commodity. He argued that treating food as merely a commodity erodes its social importance to communities and encourages food to be produced in a fashion much like other commodities. He was well aware that general public awareness through the broader dissemination of information about how the agri-food system works would play a pivotal role in the successful conversion from unsustainable agricultural techniques *back* to more sustainable ways of producing food. He had witnessed the shortcomings of monoculture through its susceptibility to infestations of pests and diseases, even with chemical applications convincing him that organic techniques were the preferred method of producing food for long-term sustainability (Conford, 2001:54).

Although Howard did not originally use the word 'organic' to describe his agricultural techniques, he did lay the foundation for a proto-definition of organic agriculture that contains three essential elements that must be present to have a safe and sustainable agri-food system: fertile soil, freshness of food products and stabilized cost (Belasco, 1989:71; Howard, 1947:28). All three of these requisites would serve as foundations for the process-based definition of organic. The factor of stabilized costs of foodstuffs proved elusive to promoters of the process-based definition because organic agriculture still had to participate in market transactions based on supply and demand. However, the agricultural techniques recorded by Howard provide a foundational basis for the process-based definition that would later come to emphasize the social and environmental goods associated with organic agriculture.

Howard's ideas became highly influential to those who believed chemical agriculture was harmful to the soil's fertility and to human health. His writings reached North America through J.I. Rodale in the 1940s. Rodale is credited as being the first person to use the term 'organic' in North America (Kuepper and Gegner, 2004). Inspired by Howard's example, Rodale decided to move from his home in New York City in 1940 to establish an experimental organic farm in Emmaus, Pennsylvania. He named the farm 'The Soil and Health Institute' in 1947 and later renamed it 'The Rodale Institute'. The Rodale Institute continues to advocate for the expansion of organic agriculture and has the mandate of 'put[ting] people in control of what they eat' (Rodale Institute, 2006).

Rodale felt that there was an intimate link between the health of the soil and the health of people. Although he incorporated many of Howard's ideas into his own activities, Rodale added his knowledge of food nutrition to his championing of organic agriculture. He promoted organic agriculture as a healthier alternative to food produced through chemical agriculture (Belasco, 1989:71). Working with Howard, Rodale experimented with organic techniques and published the results of his efforts. Publications from the Rodale Press attempted to educate readers about the environmental harm caused by chemicalized agriculture, compromised food quality and the role agribusiness played in attempting to discredit organic practices. Important publications for the Rodale Press include the *Encyclopaedia of Organic Gardening* (1959), *How to Grow Vegetables and Fruits by the Organic Method* (1959) and *The New Farm* (est. 1975) periodical, mainly geared towards organic farmers.

The Rodale Press first published *Organic Gardening and Farming* (*OGF*) in 1940 (later renamed *Organic Gardening* and relaunched as *Organic Life* in 2015) and then *Prevention* magazine in 1950. Both of these periodicals intended to educate Americans about healthful food and the importance of practicing sustainable agriculture by using 'organic' inputs. Rodale's organic agriculture magazine was originally called *Organic Gardening* (*OG*). Due to the growing number of organic farmers in North America, Rodale decided to add 'and Farming' to the title of the magazine to appeal to a wider audience. In 1970, *OGF* magazine acknowledged that people under 30 were the primary demographic driving the expansion of the market for organic food. In M.C. Goldman's piece for *OGF* on the growing market for organic products in Southern California, she describes how participation in the organic food market by the 'with-it' kids helped to get older members of society to participate in the organic movement by educating them on the economic, social and environmental ills of industrialized agriculture. As one shop-owner tells Goldman in an interview, 'a full third of our turnover now comes from the "upper-class hippie" group' (Goldman, 1970:39). Goldman frames the growing interest in organic agriculture by young people as a major source of political influence over time, as 'the influx of young people is like a blood transfusion to the health-food business, particularly to the retailers of organically-grown foods ... People under 25 will control the balance of the vote within a few years' (Goldman, 1970:40).

In 1958, *OGF*'s circulation was approximately 60 000. One year later, *Prevention* and *OGF* together had a circulation of 260 000. By 1970, *OGF* subscriptions alone grew to 650 000 (Levenstein, 1993:162). Despite the fact that *OGF* did not explicitly promote social change to accompany practicing organic farming, it was widely viewed as *the* source of information for members of the organic movement. Rodale's *OGF* in 1971 was considered by many to be the 'bible' of the movement as it provided readers with the tools necessary to participate in organic farming's alternative market, what many considered to be the true act of rebellion against the industrialized agri-food system (Greene, 1971:31).

One contributor to *The Whole Earth Catalog* (*WEC*) in 1971 declared *OGF* to be 'the most subversive publication [and that] … the whole organic movement [is] exquisitely subversive' (Norman, 1971:50).

The *WEC* regularly included reviews of organic gardening books, including Howard's *An Agricultural Testament* (1943), and a number of publications from the Rodale Press (Berry, 1969:33). Monthly publications such as the *WEC* are extremely important to understand how the process-based definition of organic was transmitted to the broader public as the movement matured in the 1970s. Gurney Norman, a contributor to *WEC*, stated that:

> organic gardeners are in the forefront of a serious effort to save the world by changing man's orientation to it, to move away from the collective, centrists, super-industrial state, toward a simpler, realer one-to-one relationship with the earth itself. Most of the current talk about 'ecology' in America is simply the noise that accompanies all fads. (Norman, 1971:50)

Wendell Berry was also concerned with the direction the environmental movement was taking in the 1970s, and urged readers of the *WEC* to practice organic agriculture based on the emerging set of environmental principles associated with the process-based definition. It was also considered a form of political protest: 'a person who is growing a garden, if he is growing it organically, is improving a piece of the world' (Berry, 1970:5). Publications like *OGF* and *WEC* were important ways of communicating the principles of organic agriculture to the broader public. When put into practice, the principles were championed as a challenge to the status quo. They served to help spread the message that organic agriculture was a principled way of producing food that rejected many of the practices used in conventional agriculture. These publications, as further discussed in Chapter 6, were fundamental to the promotion of the process-based definition of organic.

Rodale believed that the environmental benefits of promoting and maintaining biodiversity and the nutritional value of 'non-chemical' agricultural methods were fundamental to improving food quality and the health of the soil (Clunies-Ross, 1990; Rodale Institute, 2012). His ideas regarding 'regenerative agriculture' provided the process-based definition of organic with its foundational principles emphasizing environmental sustainability. Limiting dependence on non-renewable resources was key to his ideas about sustainable agriculture, and he challenged the practices of chemical farming that relied on synthetic inputs derived from fossil fuels (Berry, 1976:140). Although Rodale's work, organizations and publications did not link organic agriculture to broader social issues such as poverty or labor unionization, he recognized the intrinsic social and cultural value of the 'family farm' to rural communities and to the way food production was organized. Like Balfour and Howard before him, Rodale was critical of corporate expansion into agriculture, as it displaced many people from rural areas and made it difficult for smaller farming operations to remain economically competitive (Kuepper and Gegner, 2004).

Closely linked with Rodale was organic advocate Paul Keene. Keene and his wife Betty established Walnut Acres in 1946, the first business in the US to sell organic food through a mail order catalogue. Keene was also a long-time activist for organic techniques and published *Fear Not to Sow* in 1988, a series of essays reflecting on his experiences as an organic farmer. Like Rodale, Keene advocated for organic farming techniques and for emphasis to be placed on the ecological impacts of food production. Despite Rodale's ceaseless efforts to convert farmers from practicing conventional agriculture to experimenting with organic techniques, the financial incentives provided by government to farmers who chose to use synthetic inputs in farming proved to be too strong in the 1940s. As a result, the number of individuals practicing organic agriculture in the US and Canada remained small (Belasco, 1989:71).

The spread of organic agriculture in Canada followed much the same pattern as in the US, however the early organic movement in Canada is not as well documented. Many of the ecological and social principles that became part of the emerging organic movement in Canada originated from Europe and the US, beginning with Howard and Rodale's collective promotion of the link between the health of the soil and healthy food in the 1950s. Film-maker Christopher Chapman established one of the first organic agriculture organizations in Canada. The Canadian Organic Soil Association, later renamed the Land Fellowship, was created in Ontario in the 1950s (Hill and MacRae, 1992). Chapman, along with fellow leader Spencer Cheshire, traveled across Canada distributing information about the benefits of organic farming to the soil, the nutrient-content of food and to human and environmental health. Although organic techniques received little attention from the farming establishment in other parts of Canada, organic agriculture was welcomed in Quebec, where European immigrants had brought with them the organic techniques they practiced in their native countries (MacRae, 1990). Both the efforts of the Land Fellowship and early organic farmers in Quebec helped to establish an organic farming presence in Canada in the mid-twentieth century.

The Components of the Process-based Definition of Organic

The nascent process-based definition of organic emerged based on practitioners' dedication to finding a more sustainable way of producing food while attending to the health of the soil. After several decades of experimenting with organic techniques, mounting evidence challenged the skeptical view that organic agriculture was economically unviable and unproductive (Harter, 1973; Harrison, 1993). The three elements of the process-based definition of organic – economic viability and environmental and social sustainability – emerged from practices of early practitioners who were committed to finding alternatives to conventional agricultural practices. The principles are difficult to separate because in many ways they are interdependent, and a commitment to one element entails commitment to the others. Though what is presented here is a coherent set of principles that together make up the process-based definition of organic, it does not infer that all organic practitioners put all of these principles into practice in the same way. What the process-based definition of organic consists of is a set of principles that emerged over time, collectively reflecting a certain set of practices that organic food became associated with in the minds of the public and for many supporters of organic agriculture. The changing relationship between the organic principles and the practices of organic farming, and the ambiguity of how the principles defined the practices and *vice versa*, would influence the changing shape of the organic food sector in the latter half of the twentieth century and have a central role in the politics of organic food. This ambiguity between principles and practices would also allow for a contending definition to emerge.

  The three elements that comprise the process-based definition emerged over time with the last element, social sustainability, becoming associated with organic food in the late 1960s. All three tenets are substantive issues of process, meaning that the principles focus on the qualities of socio-economic relationships as well as the relationship between humans and the environment. Combined, the three principles prescribe that practices within the agri-food system should cultivate sustainable and healthy relationships between people, the environment and the market. The process-based definition promotes a way of producing food that is significantly different from the methods promoted by industrialized forms of agriculture and dictated by the principle of economic efficiency. However, organic agriculture in practice does not necessarily reflect the social progressiveness found in the process-based definition of organic, though organic food in the contemporary context is often associated with progressive food politics and some elements of social justice.

Economic viability

Although idealized visions of socially conscious and environmentally sound ways of producing food are indeed part of the process-based definition of organic, the definition is also tempered by the reality that organic agriculture must financially sustain those who practice it. Organic practitioners must participate in some form of market activities in order to earn a living. This element of the process-based definition of organic is the most important to organic producers selling their products for income, although a firm commitment to the environment also plays a major role in the reasons behind people's practice of organic agriculture (Abaidoo and Dickinson, 2002). The process-based definition promotes the idea that producers should be able to make a reasonable living from producing food without exposing themselves, nature or fellow humans to toxic chemicals in the process. One of the key ways of keeping organic agriculture economically viable is to keep social and economic relations that are part of the food value chain in close proximity to one another.

A value chain is defined as 'the full range of activities which are required to bring a product or service from conception, through the different phases of production (involving a combination of physical transformation and the input of various producer services), delivery to final consumers, and final disposal after use' (Kaplinsky and Morris, 2000:4). It is a relational concept that describes all of the activities firms and workers do to bring a product from its design to its end use. The concept of value chain views the production process as made up of an interconnected chain, with firms working together and exchanging information to 'add value' to a good as it moves from upstream to downstream. Value chain analysis recognizes that influence over production processes may lie in other links, such as retailers and consumers. This is especially true for the way food value chains are organized. In terms of agri-food value chains, the Ontario Ministry of Agriculture, Food and Rural Affairs (2013) offers a more precise definition:

> Agri-food value chains are designed to increase competitive advantage through collaboration in a venture that links producers, processors, marketers, food service companies, retailers and supporting groups such as shippers, research groups and suppliers. A value chain can be defined as a strategic partnership among interdependent businesses that collaborate to progressively create value for the final consumer resulting in a collective competitive advantage.

In terms of agriculture, a simplified value chain can consist of seed distributors, producers, processors, distributors, retailers and consumers. Value chain analysis is used in studies of agri-food systems to understand how linkages in production chains relate to one another and how the behavior of other links can influence the organization of resources of another (Kaplinsky, 2000). This approach to describing the activities that are part of a production process is useful to understanding the important links between social and economic aspects of organic production processes because traditionally, those involved in organic agriculture shared some common principles regarding soil health as a priority in food production and the overall rejection of practices and activities found in industrial agriculture.

In practice, value chains comprised of firms collectively subscribing to principles of the process-based definition of organic focus on reducing the use of synthetic inputs and food packaging, and prioritize soil health in the production process. The process-based definition promotes short value chains so that farmers retain as much of the exchange value of the product as possible. In the conventional agri-food system MacRae et al. show that 'globally, distributors, shippers and retailers now retain two-thirds of the economic value of food, while the farm sector (9 percent) and input sector[2] (24 percent) share the other third' (MacRae et al., 2004:27). Retaining value at the point of production is possible by limiting the number of 'links' in the chain, thereby reducing the divisions of value garnered from the sale of organic food. Therefore, farm gate sales, box schemes, farmer's markets, food cooperatives and retail outlets that purchase locally grown foods directly from the producer and cut out the 'middle man' are favored by organic practitioners because they are understood to facilitate the retention of value by the producer (Ikerd, 1999). The idea that producers should retain the exchange value of organic foods was largely built upon the earlier environmental components, which advocated soil health and an overall rejection of industrial inputs in agriculture.

Environmental sustainability
Environmental sustainability is most often associated with organic agriculture by the general public and is the only tenet that has been widely institutionalized into public policy in Canada and the US (CAN/CGSB-32.310-99; US/OFPA90, Sec. 2104 (7 USC. 6503)). Yet, the only environmental elements taken from the process-based definition of organic and included in public policy are those pertaining to practices

---

[2]　The 'farm sector' refers to the primary stages of production, such as the production of raw commodities. The 'input sector' refers to the secondary stages of production, such as food processing.

that impact the material attributes of the food product. Examples of this include the banned use of synthetic inputs and GMOs in 'certified' organic agriculture, which is discussed in greater detail in Chapter 5. The process definition considers the environmental damage caused by using synthetic inputs derived from fossil fuels and the possible future biosafety risks associated with genetically modified seeds. It largely sees both of these products of industrialization as damaging to the environment and risky to human health, while it is concerned by environmental costs that add up throughout the value chain in the form of waste and pollution from fossil fuel based inputs (for example, transportation, packaging). There are three principles of environmental sustainability in the process-based definition of organic: shorter value chains, small-scale establishments and attending to biodiversity.

The process-based definition recognizes short value chains as serving the dual purpose of keeping profits in the hands of producers while reducing the environmental impact involved in the transportation and packaging of food typically involved in longer value chains. Closer geographic proximity is prescribed by the process-based definition of organic because large distances between the points of production and the points of consumption involve an increase in transportation costs, packaging and ultimately post-consumer waste. Both are fossil fuel intensive processes. Larger geographic distances between links in the chain also increase the instances of spoilage and waste of product as it is estimated that 25 percent of food produced in the global agri-food system rots in transit (Imhoff, 1996:429).

In addition to the rejection of distancing, the process-based definition discourages reliance on fossil fuels in every part of the process by keeping farming establishments functioning on a small scale. In 1998, the United States Environmental Protection Agency (EPA) reported that the largest cause of water pollution in the US was conventional agriculture due to its reliance on mechanized inputs, fossil fuels and animal waste from intensive livestock farming (Clarke et al., 2001:14). In 2000, the EPA's National Water Quality Inventory identified industrial agriculture, namely livestock waste and field runoff, as the source of almost half of stream and river water pollution in the US, and over 40 percent of pollution in US lakes (EPA, 2000:22). Similarly, in Canada farming practices that are reliant on fossil fuels contributed 13 percent of the greenhouse gas emissions in 1996 (MacRae et al., 2004).

In an Environment Canada study looking at 20 years of data (1990–2010) on air quality and agricultural activities, livestock operations were found to be the primary source of ammonia pollution across the country (Environment Canada, 2012). Composting as a technique favored by organic practitioners intends to reduce nutrient runoff by binding it with organic matter before returning it to the field. Keeping the scale of farms small is argued to reduce the environmental footprint of agricultural practices because there is far less dependence on fossil fuel intensive technologies such as tractors and threshers to tend and harvest thousands of acres that make up large-scale operations. In regard to raising livestock, the process-based definition of organic views small-scale operations as a way to produce less overall animal waste. The waste that is produced can be used as fertilizer for crops to return nutrients to the soil. Larger-scale food production reflecting greater economies of scale encourages monoculture and produces more waste than can be recycled back into the production process.

A more recent addition to the process-based definition's commitment to environmental sustainability and biodiversity is banning the use of GMOs in organic agriculture. GMOs and other forms of biotechnology in the food system are not only viewed as possibly dangerous to human health and biodiversity, but also as technologies that create power asymmetries and relationships of dependency on agribusiness through enforced intellectual property rights over patented varieties of crops (Organic Consumers Association (OCA), 2006a; Shiva, 2000; Howard, 2000). GM seeds are considered industrial inputs because they are technological products of agribusiness. The emphasis on biodiversity in the process-based definition encourages the cultivation of what is native to one's bioregion and the preservation of 'heritage' strains of plants and animals. With industrialization of agriculture, many of the most hardy, robust species of flora and fauna were singled out to impose uniformity and standardized inputs for stages further along the value chain. An estimated 80 to 90 percent of vegetable and fruit varieties prevalent in the nineteenth century were lost by the end of the twentieth century (Henry, 2001). The process-based definition of organic rejects the use of GMOs because they reinforce genetic homogeneity, which can lead to the spread of pests that target crops of particular genotypes, as well as the concern that GM varieties may uncontrollably out-cross with non-GM crops (Kloppenburg, 1988; Kinchy, 2012). As part of the process-based defin-ition, keeping organic agriculture an environmentally sensitive form of food production requires practitioners to account for the environmental costs occurring throughout the production process. Thus, the elements of environmental sustainability are fundamental to the adherence to the process-based definition. However, keeping these substantive environ-mental values intact in practice is largely dependent upon a commitment to maintaining a link between social relations and economic relations in organic value chains.

Social sustainability

Social sustainability was the last tenet to be associated with the process-based definition of organic, but focus on social relationships that are a part of organic agricultural practices is as important as economic viability and environmental sensitivity (Shreck et al., 2006). In a number of ways, it is the most important element of the process-based definition because it focuses on the relationship between humans, the environment and the market. It sees this link as the best way of preserving the economic viability and the environmental sensitivity in organic agricultural practices (Berry, 1977). Though all practitioners of organic agriculture have not explicitly subscribed to socially progressive ways of producing food (for example, unionization of workers), the sheer nature of the labor intensiveness associated with organic agriculture demands that the link between workers and the product be considered when practicing organic agriculture. Social sustainability can be defined as including a commitment to preserving rural communities and culture (Oelhaf, 1982). This type of sustainability is not necessarily 'socially progressive', but it does attend to the importance of social relationships to agricultural practices and how they interact in the market.

Since organic agriculture in its early form did not use chemical inputs or practice monoculture, manual labor in many cases was necessary for completing various farm tasks (for example, weeding, harvesting) in a polyculture setting. Polycultural organic farms require much more human labor by producing goods that come to fruition at various times during the year. The year-round labor requirements for organic farming have been shown to reduce the cyclical unemployment of contemporary farming and the under-employment that exists in conventional agriculture (Buck et al., 1997:8). Most agricultural labor in North America is seasonal and/or casual. Seasonal labor qualifies as being employed for less than 150 days per year (Youngberg and Buttel, 1984:174). The labor requirement for a small-scale organic farm is on average 15 percent higher than the labor necessary in conventional agriculture. Because of the higher labor requirement when organic agriculture is practiced based on the process-based definition of organic, 'organic farming ... supports more jobs per hectare of farmland contributing to social stability of farm populations and rural society' (Atkins and Bowler, 2001:69; Harwood, 1984).

The 1960s saw a dramatic increase in the number of people practicing organic agriculture. The association of organic agriculture with the political causes of the counterculture in the 1960s helped to establish the general assumption that organic food is the product of a more socially just type of food production. Practitioners who attended to labor relations within organic agricultural practices reasoned that if an organic agri-food system was to be economically and environmentally sustainable, relations in the production process had to be reproduced in a manner that rejected the relationships between business, farmers and the state that were deemed to be exploitive. The practice that makes organic agriculture socially sustainable is the small scale of production processes, which is also a way of putting environmental principles into practice. Keeping ownership of land disaggregated among families or small groups of people aids rural communities in economically sustaining themselves and helps keep decision-making over resources related to the production process and the surrounding environment in the hands of those who live and work in rural areas. Therefore, valuing the human labor involved in the organic production process acknowledges the human and environmental elements that go into organic production. As argued by Tad Mutersbaugh in his study of organic coffee growers, the local, grassroots nature of organic agriculture is what has made it sustainable agriculture (2002:1167).

The process-based definition of organic is premised on mutually dependent types of sustainabilities: economic and environmental sustainability, both of which encourage a commitment to social sustainability. By incorporating these principles into practice, organic practitioners and supporters of the process-based definition of organic demonstrate their commitment to social and environmental sustainability. What sets the process-based definition apart from other ways of producing food is its integration of social and environmental costs and its attempt to decrease the negative outcomes of agriculture on the environment and various human relations such as labor. By including and seeking to reduce the costs externalized through industrializing processes, and valuing the inclusion of organic principles in practice, the process-based definition provides an alternative for those who believe that externalizing the costs of production and ignoring the importance of social to food production is unsustainable and unethical.

Because the process-based definition of organic emerged as a challenge to relationships found in the conventional agricultural sector, its principles oppose many of the practices and relationships found in conventional agriculture. This makes it difficult for large-scale corporate actors to use conventional agricultural practices while adhering to the process-based definition of organic. Yet, because the principles found in the process-based definition of organic were not enforceable or evenly practiced by organic practitioners, corporate actors seeking to enter the organic food sector were able to employ the principles they wished, while not ignoring others. As regulatory standards for organic production emerged in Canada and the US that defined 'certified' organic foods as foods produced without synthetic inputs (including GMOs), the process-based definition of organic was no longer the only guiding set of principles for organic practitioners (discussed in more detail in Chapter 4). As organic food's association with 'health food' became stronger and the general public's interest in health food increased, so too did conventional corporate interests in investing in the organic food sector. The growing popularity of organic food in the mainstream was more about how the food itself could impact individual human health, and less about how organic principles when applied to agriculture improve soil quality, reduce off-farm inputs and contribute to social and economic sustainability of organic practitioners.

THE PRODUCT-BASED DEFINITION OF ORGANIC

The 1980s signaled a new approach to organic agriculture, reflecting something quite different from the radical political ideals of the 1960s. Whereas the process-based definition of organic emerged from the development of agricultural practices that were alternatives to conventional agricultural practices and relationships, the product-based definition does not reject conventional practices and relationships in the same way. This approach to organic emerged in the 1980s, and, contrary to the process-based definition, focuses on the production processes that directly impact the material qualities of an organic food product. In an article published in the journal *Science* in 1980, the author claims that organic farming had finally become 'legitimate' (Carter, 1980). The piece explains how organic farmers (of the day) had successfully distanced themselves from the reputation as a 'back-to-nature romantic left over from the 1960s' and were now part of the mainstream (Carter, 1980:254). Also in 1980 the USDA estimated that there were 24 000 organic producers in the US, partially due to its growing acceptance by the public as a legitimate practice that produced healthier and safer food than the conventional agri-food system. This led to growing consumption of organic food by mainstream consumers (*OG*, 1989:43).[3]

The growing popularity of organic is reflected in the increasing circulation of health-oriented magazines like *OGF*. In 1980, the magazine's circulation grew to 1.3 million and *Prevention*'s circulation was approximately 2.4 million (*Business Week*, 1980:85). The focus of *OGF*

---

[3]   Similar figures for Canada are not available.

began to change as the Rodale Press sought broader readership, reflecting some organic advocates' desire to appeal to mainstream consumers' increasing interest in health foods. J.I. Rodale's son Robert wanted his family's magazine to reach out to mainstream consumers concerned with health, nutrition and food quality. Interviewed for a *Business Week* article, Robert Rodale describes the new marketing strategy for *OGF* which focused on attracting big-name advertisers beyond the vitamin supplement and health food companies of the magazine's past. The Rodale Press planned to expand the distribution of its publications beyond health food stores into mainstream bookstores to meet growing consumer demand (*Business Week*, 1980). Although President of the Rodale Institute Robert J. Teufel said that the institute still subscribed to the ideas of 'planned growth', he was also eager to expand *OGF*'s readership: 'we are still not reaching enough of the sensitive part of the population. There is a broadened national interest in the health field and survival. If we don't fill that gap, someone else will, and they won't do it as well.' (*Business Week*, 1980:88) Although the Rodale Press was never overly interested in participating in political activism associated with some organic practitioners, the shift towards the mainstream by the Rodale Press signaled a significant change in how organic food was viewed by the general public, and how the definition of organic began to change as it grew in popularity.

Surging public interest in organic food was grounded in the growing list of documented food safety issues emerging from the conventional agri-food system including the discovery of grapes found in the US agri-food chain contaminated with cyanide (Buttel and LaRamee, 1991:164). A number of other food safety scares across North America gave credence to the organic movement's claims that the industrialized agri-food system was to blame for the rising instances of food safety issues. Food policy scholar Tim Lang labels the period between 1980 and 2000 in post-industrial countries as a time of 'public crisis' where concerns about pesticide residues on food, unnecessary additives, the role food plays in degenerative diseases and food contamination cases made the public seriously question the safety of the conventional agri-food system (Lang, 2004:28). For the organic movement, the public crisis in the capacity of the state to assure food safety served as an opportunity to restate the benefits of organic farming and attract new supporters who viewed organic food as a 'healthy' alternative to conventional fare.

The arguments made by organic practitioners that organic food was safer and healthier than conventionally produced food helped the organic movement gain momentum in the mainstream in the 1980s. The association between organic food and health was more widely publicized than the promotion of organic agriculture as a way to manage soil health because mainstream consumers were more interested in the benefits of organic food to their own health. The increased popularity of organic food as healthier and more natural than conventional food helped lay the groundwork for an alternative vision of organic agriculture. A way of promoting organic agriculture to the broader public was to appeal to consumers' desire to improve their 'quality of life' through the purchase of organic food for its perceived healthful qualities.

But how did an alternative definition of organic emerge, that did not include many of the principles associated with the process-based definition of organic? The agri-food system underwent radical change in the 1980s in terms of how production processes were organized, which allowed for a type of food production that utilized conventional practices while producing 'certified' organic foods. The two processes, described as 'appropriationism' and 'substitutionalism', helped to restructure the agri-food system to reduce costs where possible and to maximize profit. Appropriationism, as it applies to agriculture, is 'the transformation of agricultural production processes into industrial activities that minimize constraints to profit accumulation imposed by agriculture's basis in biological cycles' (Gillon, 2011a). An example of an input cost in agricultural production processes that can be reduced through appropriationism is human labor. It can be replaced with machinery through industrializing processes and the use of industrial technologies. A related strategy used to reduce overhead costs is 'substitutionalism'. Substitutionalism is defined as 'efforts to reduce agricultural products to industrial inputs for food manufacture or to entirely replace farm-based products with industrially produced substitutes' (Gillon, 2011b). Replacing inputs with cheaper varieties indistinguishable in end-product evaluation, such as iodized salt replacing more expensive sea salt as an ingredient, is a form of substitutionalism. These strategies describe the processes of industrialized agriculture and help to explain how agriculture and the rise of certain technologies like food preservatives and refrigeration radically changed how food was produced and processed (Goodman, Sorj and Wilkinson, 1987). The dual concepts are widely used to describe the changes experienced in (post-) industrial societies in North America, Europe and Latin America as methods of mechanization and industrialization were used to transform the agri-food system (Mann, 1990; Bonanno et al., 1994; Alteri and Rosset, 1997). Cutting input and overhead costs can be further achieved by lengthening value chains so that links are located in jurisdictions with the most favorable labor conditions (for example, low union density, high rates of unemployment, lower employment standards), finding cheaper suppliers of ingredients or linking firms that specialize in certain aspects of the production process to achieve economies of scale.

The focus on meeting consumer demand for 'healthy' organic food, in conjunction with the lack of a universal definition of organic and an inconsistent implementation of the process-based principles, created an ambiguous situation where it was unclear as to what principles had to be put into practice in order for the production method to be considered organic. Because of this lack of clarity between principles and practices, corporate actors employing business practices found in the conventional agri-food system had few restrictions placed on their entry in the organic food market. There was nothing in the process-based set of principles that explicitly restricted corporate involvement because of the assumption that practitioners were drawn to organic agriculture because of its principled practices. With no clearly defined boundaries for organic, corporations were able to define organic in a way that facilitated their business practices. The 'product-based' definition is premised on a set of practices geared towards meeting the demands of consumers for food free of synthetic inputs while capitalizing on the association of organic food with food safety. It is not necessarily based on the principles associated with the process-based definition, though the end product is technically organic (Synovate and Winam, 2003; Hallam, 2003:185; Walnut Acres, 2005).

The product-based definition of organic does not necessarily consider how organic food is distributed, the scale of organic agricultural establishments or the social relations that are part of the production process. It does not necessarily reject the practices and normative behaviors associated with industrialized agriculture. According to this approach, a good can still be considered certified organic even if its ingredients are cultivated using industrial machinery and fossil fuels. While expanding the markets for organic products encourages the reduction of synthetic pesticides, herbicides, fungicides and GMOs in the agri-food system, which meets some of the environmental objectives associated with the process-based definition, the product-based definition does not include any prescriptions for how organic foods should be distributed in an environmentally sustainable way. Adhering to the product-based definition does not discourage the sale of organic products through conventional distributional channels like supermarket chains or global value chains (Thomson, 1998). Where the ambiguity between principle and practice becomes problematic is when organic foods produced in accordance to the product-based definition are promoted and continue to associate with the more substantive principles of the process-based definition, which is associated with all organic foods carrying a certified organic label.

For proponents of this definition, the positive outcomes of main-streaming organic foods include the increase in land devoted to organic production methods, while making organic options available to a wider demographic of consumers. The expansion of the product-based definition of organic has led to a major shift in where organic foods are available and purchased. Since 2000, consumers in the US purchased almost half of their organic products through conventional supermarkets, while in Canada, in 2003, between 45–50 percent of organic products were sold through conventional retailers (Murdoch and Miele, 1999:478; Dimitri and Greene, 2002:2; Macey, 2004:25). This figure has changed little in the past seven years according to the Organic Trade Association (OTA), with Canadians purchasing 41 percent of their organic products from conventional grocery stores in 2010 (Lazarus, 2010).

Proponents of the product-based definition claim that the overall increase in the consumption of organic food is environmentally beneficial because it means that fewer synthetic chemicals and GMOs are released into the environment (OTA, 2006). This is a benefit to those who oppose biotechnology in the food system and the environment, and contributes to the decreased use of synthetic agricultural inputs, both aspects of the environmental sustainability principle included in the process-based definition. Yet, the political implications of applying the product-based definition to practices found in organic value chains also legitimizes the conventional approach to how food products are evaluated, which is based on equivalence and sameness; for example, the concept of 'like' products enshrined in international trade agreements (see Chapter 5). A Canadian organic Spartan apple grown on a large-scale orchard and an American organic Spartan apple from a small-scale orchard are defined as equivalent products as long as they both do not contain GMOs or synthetic inputs (and they are both determined to be the Spartan variety of apple). No other elements make either apple organic other than the absence of synthetic inputs; not the length of the value chain, not where it was produced, nor whether one apple is a product of monoculture and the other is not. The 'like' product concept is also included in formal regulations for certified organic production in Canada and the US (though some differences in regulations exist). The formalized definition of organic that includes product equivalence left out some of the important parts of the social, economic and environmental elements that advocates and supporters of a process-based definition of organic agriculture view as fundamental to organic practices.

Despite the environmental damage attributed to the transportation necessary for transnational value chains to function, the product-based definition of organic justifies the expansion of distances between links in organic value chains as essential to meet consumer demand for organic foods or ingredients that are not grown locally. Because in many cases localized value chains cannot generate sufficient quantities of organic products that are demanded (Macey, 2004), the instrumental approach to organic increases the variety and supply of organic goods to consumers. Global transportation networks, it is argued, are necessary to meet consumer demand for organic products despite their reliance upon fossil fuels (Bentley and Barker, 2005).

Producing organic products also has a valuable market function for corporations that do not necessarily subscribe to organic principles included in the process-based definition. Though scholarly and media attention has revealed how corporations producing organic products use images and language in marketing campaigns that allude to the process-based definition, the public perception that organic foods are produced in accordance to the principles found in the process-based definition continues to be strong. The association with a more socially conscious form of food production, whether or not it is reality, is a valuable marketing tool. In her book *The Empire of Capital*, Ellen Meiksins Wood addresses the notion of 'ethical' corporations and corporate 'social responsibility'. She states that, 'even the most "responsible" corporation cannot escape [the] compulsions [of putting exchange value before use value], but must follow the laws of the market in order to survive – which inevitably means putting profit above all other considerations, with all its wasteful and destructive consequences' (Wood, 1999:14). The price a firm can receive for a good will always be prioritized over the value of a good. Since organic products have always been subject to exchange value (production requires consumer's willingness to pay), it has not been too difficult to incorporate them into the conventional arena of exchange value such as global trade. Today, state-sanctioned certified organic labels reflect the product-based definition of organic, referring only to the attributes found in the material end product. It assures consumers that the product carrying a certified organic label was not produced with the use of synthetic chemicals or GMOs. It tells consumers very little about whether the product was produced in accordance with the process-based principles often associated with the certified organic label.

CONCLUSION

Early practitioners of organic techniques were aware of rising large-scale corporate interest in organic agriculture as early as the 1970s (Belasco, 1989:99), but it is difficult to determine whether they anticipated the extent to which the involvement of corporations in the organic sector and rising mainstream consumer demand would present an ideological challenge to the viability of an understanding of organic that prioritized the use-value of the process, in addition to its exchange value. Many early advocates following Howard and Rodale's ideas wanted organic techniques to gain popularity and for people to purchase more organic products. At the same time, they wanted organic agriculture to be practiced based on principles that pay significant attention to what happens throughout the production process as an important determinant of the value of organic food that differentiates it from conventional fare.

Examining the historical trajectory of the process-based definition of organic and the rise of the product-based definition of organic shows that as organic food has gained in popularity and corporations continue to enter the sector, the normative understandings of what constitutes 'organic' distanced from the developing institutionalized definitions in regulatory frameworks. The weak links between principles and practices when the organic movement was still in its infancy created a gray area that allowed for the product-based definition to take shape, and would later influence standardized regulations for organic agriculture and certified organic foods. Two sets of organic principles now exist and cater to two different worldviews (Table 2.1). One sees the current global agri-food system that is based on the principles of economic efficiency and economies of scale as environmentally, socially and economically unsustainable and unjust. This worldview is an important foundation for the process-based definition of organic. The other approach, though including some of the tenets of process-based principles, tends to prioritize the expansion of the production of and markets for organic foods. Although not all organic practitioners strictly put either definition into practice, the typology introduced in this chapter provides a clearer understanding of two predominant definitions of organic that developed over time and are put into practice in a variety of different ways.

Table 2.1   Contending definitions of organic

	Shared values	Process-based	Product-based
World view	promote the expansion of organic agriculture	multifunctional social, environmental and economic relations are necessarily interconnected	converges with liberal market principles; open markets are the best way to spread organic agriculture
Economic elements	price premium	value stays with producer; short, localized value chains; producer-centric	expand markets; maximize efficiency; economies of scale; consumer-centric
Environmental elements	no GMOs or synthetic inputs	polyculture/ biodiversity; small-scale production	eco-packaging
Social elements	N/A	grassroots decision-making; producer-centred; sustainable labor practices	increased consumer access to product

3. Business as usual? Conventional corporate strategies in the organic food sector

> We no longer position ourselves as the alternative ...
> We are mainstream.
>
> vice-president of marketing,
> Whole Foods Market (1987)[1]

INTRODUCTION

The year span of 1999–2000 proved to be a landmark period for corporate investment in the organic food industry in Canada and the US. During this time there were 25 major acquisitions of organic firms in North America, the highest number in a 12-month period to date. This pace continued throughout the first decade of the 2000s and, despite the global economic downturn in 2008, corporate interest in the organic food sector has not waned. In 2010, there were ten announced mergers and acquisitions of organic and natural foods companies in the US. Almost every major agri-food corporation operating in Canada and the US now holds a financial stake in the organic food industry through mergers, acquisitions, partial equity or brand introduction (see Appendix 1). This chapter traces how corporate strategies more often associated with the practices of agribusiness were introduced and replicated in the organic food sectors in Canada and the US. It shows that a key element to the introduction of these behaviors was the successful reorganization of production processes in the organic sector that hinged on the promotion and acceptance of the product-based definition of organic. It begins by discussing the economic reorganization that happened in the conventional agri-food sector in the 1980s to explain the emergence of these strategies intended to capture market share. It then analyses how they were applied

---

[1] Fleur Hedden, vice-president of marketing, Whole Foods Market (1987). Quoted in Burros, 1989.

in the organic sector by highlighting the activities of three of the largest organic and natural food corporations functioning in Canada and the US: Hain Celestial, United Natural Foods and Whole Foods Market. The chapter concludes with a discussion of the implications of restructuring the organic value chain in North America for the viability of process-based definition of organic.

RESTRUCTURING THE AGRI-FOOD SECTOR

The way food was produced in Canada and the US rapidly changed in the 1980s (Wolfe, 1998), with major implications for the shape and development of the organic food industry emerging in the 1990s. To understand how conventional agri-food corporations became stakeholders in the organic food industry it is important to discuss the major structural changes in the agriculture sector that preceded its growth. Governments in both the US and Canada have supported their domestic agricultural sectors since the early twentieth century, treating agriculture as a 'domestic issue' of national security. But the economic recessions that plagued the global economy from the late 1970s onwards ushered in a different approach to managing the agri-food system in Canada and the US. Market principles used to guide other sectors of the economy were increasingly applied to the agri-food system. The acceptance of liberal market principles as the best way to manage all economic sectors gained support from governments of both countries to dismantle the distinctive interventionist agricultural policies that had been part of national economic strategies since the end of World War II (McBride, 2001; Clapp and Fuchs, 2009).

One sign of restructuring in the agricultural sectors in both Canada and the US was the changing organization of farming establishments. Recent decades have seen a steady decline in the number of individually owned and managed farms in Canada and the US, what is commonly referred to as 'the family farm' traditionally owned and operated by a farmer and his/her family who lives on the farming property. Between 1961 and 1971 the number of individually owned farms in Canada declined by 115 000, while in the US between 1964 and 1978 the number of such farms declined by 900 000 (Statistics Canada, Series M12-22; USDA, 2002b). Table 3.1 shows that in both Canada and the US since the early 1980s, the total number of farms has steadily declined, with the average area in acres increasing in Canada and remaining fairly stable in the US.

Table 3.1 Total number of conventional farms average and area (in hectares); total land under cultivation (1981–2011)

Canada	1981	1986	1991	1996	2001	2006	2011
Total number of farms	318 361	293 089	280 043	276 548	246 923	229 373	205 730
Average area	207	231	242	246	273	294	315
Total land under cultivation	N/A	67 825 757	67 753 700	68 054 956	67 502 466	67 586 747	64 812 732

United States	1982	1987	1992	1997*	2002*	2007	2010
Total number of farms	2 240 976	2 087 759	1 925 300	1 911 859	2 128 982	2 076 000	2 200 930
Average area	178	187	199	197	178	182	169
Total land under cultivation	399 424 000	390 166 000	382 832 000	386 070 000	379 959 000	376 730 000	372 310 790

Notes: 1 hectare = 2.471 acres.

*Adjusted for coverage.

Sources: Statistics Canada (Census), 2004, 2006, 2011; USDA, 2004, USCB, 2012.

Between 2001 and 2006, the number of Canadian farms declined by 7.1 percent (Statistics Canada, 2007) and between 2006 and 2011 by 11.5 percent. The decline was smaller in the US between 2001 and 2005 with the number of farms declining by 2.5 percent (Hoppe and Banker, 2006), with a further 5.7 percent decline between 2007 and 2010 (United States Census Bureau (USCB), 2012). The decline in the number of individually owned farms in the US, however, is accompanied by a slight decline in land devoted to agriculture. In Canada, the land devoted to agricultural production remained between 67 and 68 million hectares (ha) from 1996 to 2006 (Statistics Canada, 2007) with a dip to 64 million ha in 2011. In the US, the land devoted to agriculture between 1997 and 2002 declined from approximately 386 million to 379 million ha and then fell to 372 million ha in 2010.

There are some difficulties in comparing the changing structure of agriculture in Canada and the US because both countries use different typologies in organizing farm data. For example, the definition of family farms is different in Canada and the US. The primary difference in how the family farm is defined is based on income derived from farming operations. 'Small family farms' in the US are defined with sales less than $250 000 (USD) per year, while under one component of Canada's definition, a family farm (of the 'very large business-oriented' variety) can earn up to $500 000 (CAD) in revenue (Hoppe et al., 2004:85). Hoppe et al. (2004) describe the patterns of similarity and difference between farms in Canada and the US, and how they have changed structurally over the last few decades. Some examples of general similarities between farms in Canada and the US include the high concentration of income among larger farms, whereas smaller farms run by family units rely on off-farm income to supplement what income is generated on the farm. As for differences, Hoppe et al. (2004:84) state the prevalence of 'retirement and lifestyle' grouping of farms in the US. Nearly half of the farms fall into this category. Operators in this category attain much of their income from off-farm sources. The number of non-family farms (according to the Canadian farm typology) is also much smaller in Canada than in the US, whereas the number of 'large business-focused' farms is the largest single group in Canada. Interestingly, the report points to the respective tax codes as partial explanations of the differences in changing patterns of farm structure in Canada and the US. The US tax code allows for almost unlimited writing off farm losses against other income, whereas Canada has a maximum of $8500 (CAD) (increased to $8750 since the report was published). Further, Canada's marginal tax rates create incentives for farmers to reinvest in their farms as opposed to converting income to capital gains. This, Hoppe et al. argue, is one explanation for a greater number of large business-focused farms in Canada (2004:88).

The shift from a large number of small-scale family farms to a smaller number of larger-scale farming operations led to a major change rural employment. The shift in the agricultural labor market has been characterized as 'labour displacement and replacement', meaning the replacement of uncommodified family labor with wage labor (Friedmann, 1978). Some family farms converted to managerial styles of 'lean' production to compete with corporate farms, which often result in the need for off-farm employment for farmers and their families. Much of the employment in agriculture situated on small-scale farms in the early 2000s according to Statistics Canada has shifted from the production segment of agriculture, or what is labeled 'the agriculture group' (for example, farm workers, veterinarians), to the service segment or the agri-food group (workers in food retail, processing food and beverage services) (Keith, 2003:4). By 1996, 77 percent of all employment in the Canadian agri-food sector was in the agri-food group, showing a significant shift in employment from primary stages of the agri-food value chain to processing and distribution stages (Keith, 2003:7). Statistics Canada reported in 2002 that while farm employment has fallen (and continued to fall between 2006 and 2011 by 10 percent) output has not, yet farmers have not seen an increase in profits since 1996 (Statistics Canada, 2002b). The profits made from sales of agricultural commodities are highly concentrated in the Canadian sector. In 2001, the largest farms in Canada (5 percent of the total) earned one-third of the total farm revenues (Qualman and Wiebe, 2003:15). A 2005 USDA report on agriculture found that large and very large farms made up 9 percent of the total number of farms but account for 73 percent of farms sales (Hoppe and Banker, 2006).

Labor shortages in both Canada and the US have contributed to the creation of government-sanctioned programs designed to bring foreign workers into both countries to perform manual labor on farms. Due to the difference in size of agricultural sectors and types of crops grown, the labor needs on farms in Canada and the US differ significantly. The labor on farms that cannot be performed by machines is manual labor, which is sometimes performed by temporary seasonal workers (Smith et al., 2002:48). Agricultural worker programs exist in both Canada and the US, though there is a large contingent of undocumented agricultural workers in the US not covered by the provisions set out in the US program. The Seasonal Agricultural Worker Programme (SAWP) is administered by the Department of Human Resources and Skills Development Canada (HRSDC).

Beginning in 1966, the SAWP originally focused on bringing temporary seasonal workers from Jamaica and other Caribbean countries, but has since expanded to Mexico in order to fill labor gaps in the agricultural sector. As of 2013, SAWP operates in Ontario, New Brunswick, Alberta, Quebec, Prince Edward Island and Nova Scotia. These provinces receive 90 percent of the workers through SAWP. British Columbia (BC) joined SAWP in 2004. Canada had over 18 000 foreign migrant workers in the agricultural sector in 2002 (Justica, 2013). This number steadily rose between 2006 and 2009 from over 24 000 to almost 28 000 across the above listed provinces (HRSDC, 2010). Workers primarily labor in fields, orchards and greenhouses harvesting vegetables, fruits and inedibles, such as tobacco, sod and flowers. Workers through the SAWP receive an hourly wage of $8 per hour with some exceptions and are sent back to their native country once their contract is completed (usually over a three-month period).

The H-2A guest worker program in the US is set up to fill labor gaps in agriculture. Before an employer is allowed to bring in workers from outside the US, she/he must demonstrate that there are no US workers available, able or willing to apply to the Department of Labor for a temporary worker certification to fill the positions. There are several protections for workers (reinstated by the Obama administration after the previous Bush administration amended key components of the program) covering wages, treatment, and housing and safety standards. Total figures of actual number of workers brought into the US for the purpose of agricultural labor on a yearly basis are difficult to locate. Various governmental agencies (Dept. of Homeland Security, Dept. of Labor, Dept. of the State) collect data on temporary seasonal workers, but do not publicly release figures (Farmworker Justice, 2014). The number of temporary workers brought into the US for agricultural labor purposes is also obscured by the presence of a large number of undocumented workers laboring on farms throughout the US. It is estimated that anywhere from 50 to 80 percent of all agricultural workers are undocumented in the US (Getz, Brown and Shreck, 2008:484). The staggering number of undocumented agricultural workers in the US is a well-known phenomenon. Because of US immigration restrictions, and poverty and conflict in home countries, people from around the world end up in the US as undocumented agricultural workers to attempt to earn an income. In 2004, it was estimated that there were over two million foreign migrant farm workers in the US (Ahn, Moore and Parker, 2004:1). Because they are undocumented, and not counted through conventional guest worker programs, it is difficult to estimate how many undocumented workers there are at any given moment in the US. In California, it is estimated that there are approximately 800 000 agricultural workers every year because of its large agricultural sector. It is not known what percentage of the workers coming from Mexico is undocumented.

The organic agricultural sector is not immune from the realities of labor shortages and other economic realities of agriculture in North America. The use of undocumented workers in organic agriculture has been observed and as Getz, Brown and Shreck (2008) argue, in some cases organic farmers in California have opposed the strengthening of labor rights for agricultural workers because of their perception that agribusiness interests are a core component of labor policies. Farmers also worry that higher wages could affect their competitiveness in the market. Despite the common association of organic production systems with progressive social politics, in reality organic agriculture practices do not always reflect the principles of a process-based definition that considers the social relations in the production process.

Once populated with millions of small-scale family farms, the Canadian and American rural landscapes are replaced with larger-scale farms that do not require the type of skilled labor that was vital to rural farming communities for almost a century. Some organic practitioners are opposed to the 'de-skilling' of the agricultural workforce and other broader social and economic changes in the agricultural sector, and have tried to develop organic agriculture as an alternative to this model of displacement and replacement in the agri-food system. Certain segments within the contemporary organic movement continue to oppose the consolidation and labor deskilling of the agricultural sector such as the National Farmers Union (NFU). The NFU represents farmers who practice organic and conventional agriculture (mostly small-scale) in Canada and the US, and is concerned with the value of agricultural goods moving further away from farmers and to other downstream points in the value chain. The NFU is very critical of the shift in agricultural production from small-scale farms to large-scale farms, charging, 'current government policy, in effect if not intent, is often no more than the promotion of these corporations' agendas ... [which] conflict with the best interests of farmers, farm families, rural communities, as well as with those of consumers' (NFU, 2013). The price producers receive from retail food sales has steadily declined in the past four decades (Table 3.2) making farming an increasingly difficult occupation for those wishing to practice agriculture on a small scale.

Table 3.2   *Farmers' share of the retail food dollar (USD) in percent**

Select commodities	1970	1993	2000	2010
Cereal and baked products	16	7	5	3
Processed fruits and vegetables	19	19	17	–
Beef	64	56	44	22
Pork products	51	37	30	13
White bread	9	5	5	4
Market basket of food products	37	26	20	16

*Figures are rounded.

Source: Heffernan and Hendrickson, 2002; NFU, 2010.

CONVENTIONAL CORPORATE STRATEGIES IN THE ORGANIC SECTOR

To understand the development of conventional corporate strategies in the organic sector in the 1980s it is helpful to explore some of the factors that contributed to the growing consumer demand for organic foods in the 1960s. As far back as 1969, a poll conducted by National Analysts Inc. on 'Americans' health practices and opinions' indicated that a majority of Americans believed foods grown using natural techniques were more nutritious than those grown with chemicals (Levenstein, 1993:163). Consumers were also increasingly demanding traceable, healthier, cleaner, safer and more 'natural' products from the agri-food sector (Murdoch and Miele, 1999:478; Levenstein, 1993:162). Documenting the rise in demand for organic food began as early as 1972. There are a notable number of articles in publications such as *Food Technology*, *Newsweek* and *Business Week* from the early 1970s documenting the increased prevalence of organic food in the mainstream market (Hewitt, 1970; Goldman, 1970; Greene, 1971; White, 1972; Marshall, 1974; McBean and Speckmann, 1974). When discussed, the category of organic food was often conflated with 'natural food' (more generally known as 'health food'). An article by Wolnak published in the *Food Drug Cosmetics Law Journal* characterized health food (including organic) as 'fad food'. He estimated that the US market for health food in 1971 was worth between \$250–300 million (USD), whereas he cites another estimate market value of $500 million (USD) in 1972 (Wolnak, 1972:455). By 1975, the value was projected to be 'easily ... one billion plus dollar market', and three billion by 1980. Rising consumer demand for food produced without chemicals was a major driving force behind the expansion of the health food market that was partially driven by the documented cases of food safety hazards in the conventional agri-food system.

Trading on these concerns, as Nestle argues in *Food Politics,* conventional agri-food corporations aggressively engaged in the development and promotion of healthier food products to 'overcome the infamously slow growth of the food industry as a whole' in the 1980s (Nestle, 2003:318). The corporate drive to invest in the natural and health food market was facilitated by the rise in cases of tainted and unsafe food coming out of the industrialized agri-food system and growing consumer fears over food safety. Food safety scares throughout the agri-food system called into question the industrializing processes that dominated food production. Events such as the E. coli crisis of the 1980s, the 1989 'Alar food scare'[2] and botulism cases, and numerous food recalls have highlighted the perils of the globalized agri-food system, contibuting to the growing market for organic and natural products. Although there continues to be controversy over whether organic food is indeed safer and more nutritious than conventional food, agri-food corporations are eager to capitalize on consumer fears about food safety and food quality, and to meet the growing demand for organic products.

Organic food has been integrated into the conventional agri-food sector in three distinct ways, as identified by Howard (2006): mergers and acquisitions by agri-food companies; the forging of relationships built through strategic alliances; and the introduction of organic brand name products. The following sections use Howard's typology to examine the changes that have helped to tranform the organic sector in Canada and the US into a multi-billon dollar industry that continues to grow.

Mergers and Acquisitions: Vertical and Horizontal Integration

Mergers and acquisitions in the agri-food sector have always existed, but corporations began to consolidate various agri-food sectors in the 1980s like never before. Market actors faced with pressures to liberalize agricultural production were forced to either 'get big, or get out' (Berry,

---

[2]   The Alar food scare occurred when high levels of the chemical were found in apples circulating through the US food system.

1977:41). Heffernan and Constance have labeled the corporate consolidation of ownership in the agri-food sector in the 1980s as 'merger mania' (1994:39). Mergers create corporate conglomerates consisting of a number of smaller firms with vast amounts of capital to invest to secure market shares and reduce competition. It is estimated that the production of over 80 percent of all value-added food products in the early 2000s was controlled by 100 firms worldwide (Lang, 2003:18). Though mergers began to play a more important role in the organic food sector in the 1980s (such as in the case of Whole Foods Market and United Natural Foods), firm acquisitions occurred at a much higher rate from the mid-1990s until the end of the 2000s. According to a *Forbes Magazine* article in 2010, the top five food companies (PepsiCo, Dole, General Mills, Nestlé and Kraft Foods) dominating the US conventional agri-food market heavily invested in the organic food sector through mergers or acquisitions, beginning with Kraft's purchase of Canadian company LifeStream in 1981 (*Forbes Magazine*, 2010). LifeStream was established by Arran Stephens (who would later establish BC-based Nature's Path) in 1971. LifeStream's sales that year were $9 million (CAD), making it one of the most profitable natural food lines in Canada (Nature's Path, 2013).

The acquisition of firms producing organic food by conventional agri-food companies, which began in the early 1980s, has taken the form of either vertical and horizontal integration (see Appendix 1). Vertical integration occurs when one firm invests in another firm that specializes in other stages of the value chain; for example, a tomato processor acquiring tomato farms. Through vertical integration, outputs from one stage of the value chain can serve as inputs for another stage and there is much more control over the quality of the raw material supply. Vertical integration brings all the production processes in a value chain under centralized management, thereby reducing uncertainty, streamlining production through reaching economies of scale and lowering transaction costs while limiting competition (Cohn, 2002:333). Vertical integration has occurred to varying degrees in the organic food sector, and vertically integrated value chains include individual farmers, cooperatives, wholesalers, processors and retailers (DeLind, 2000:202). An example of the vertical integration of organic firms into the ownership structure of a conventional processor is US-based M&M Mars' purchase of the organic seed company Seeds of Change in 1997. In the case of M&M Mars, the acquisition of Seeds of Change was not meant to supply its other subsidiaries but to gain market share in a primary stage of the organic value chain. M&M Mars did not intend to convert its product lines to conform to organic regulatory standards, so investing in Seeds of Change is best understood as a strategy to profit from the sales of organic seeds without investing in the other elements of the organic food sector. As of 2015, the acquisition of Seeds of Change is the sole effort by M&M Mars to incorporate organic firms into the company.

One of the best recent examples of a transnational corporation (TNC) attempting to vertically integrate organic firms is Canada-based SunOpta's purchasing of businesses in both Canada and the US. SunOpta, previously named Stake Technologies, is the largest provider of organic soy in North America and is also involved in producing organic corn. It supplies organic feed to organic poultry producers in both Canada and the US, and is the only large-scale supplier of organic chickens in Canada (Sligh and Christman, 2003:18). Since 2002, SunOpta has acquired three Canadian organic processors in an attempt to consolidate control over that segment of the organic sector: Simply Organic, Organic Kitchen, and Kettle Valley Dried Fruit Ltd (SunOpta, 2004:44–5). Although SunOpta mainly concentrates in primary processing it has also expanded its portfolio to organic distributors in Canada. SunOpta has acquired four Canadian organic distributors: Distribue-Vie (Que), Snapdragon Natural Foods Inc., Wild West Organic Harvest and Pro-Organics. All of the organic processors SunOpta purchased were independently owned firms. Like other large-scale corporations in the US, SunOpta purchases smaller firms in the organic sector to consolidate ownership over a significant segment of the Canadian organic sector. According to SunOpta's 2003 annual report, 'SunOpta has become the largest distributor of organic fresh foods in Canada and is quickly reaching its objective of becoming the first national distributor, integrated from organic fresh foods, to grocery, to dairy and dairy alternatives' (SunOpta, 2004:14). SunOpta has achieved this through its internal growth strategy, which includes 'aggressive acquisitions' of many small-scale Canadian companies (SunOpta, 2004:25). Through its acquisition of organic firms at various stages of the value chain, SunOpta has vertically integrated a number of sub-sectors, such as organic soy, and continues to have the mandate of further vertical integration in Canada. For example in 2008, SunOpta acquired Amsterdam-based Tradin' which is a supplier of organic fruit juices, frozen fruits and vegetables and other partially processed foodstuffs.

In 2010, SunOpta was purchased by US-based distributor United Natural Foods (UNF). UNF has vertically integrated SunOpta into its business strategy, making UNF the largest distributor of organic and natural foods in North America. UNF itself began as a merger between Cornucopia Natural Foods and Mountain People's Warehouse in 1996. Since then, UNF has acquired nine organic and natural food distributors across the US, some of which were previously cooperatives (Howard, 2009b:101). By 2002, it rivaled Tree for Life as the biggest organic and natural food distributor in the US. Since 2002, it has steadily acquired organic, natural and specialty food distributors. In 2011, UNF was one of the five largest organic/natural food corporations with subsidiaries in Canada and the US (Capstone Partners, 2012:8).

Vertical integration also takes place at the sub-state level. In California, a number of large-scale organic farming establishments vertically integrated from the 1990s onward. As Pollan reported in his *New York Times* article in 2001, 'mega-farms' (like Earthbound Farm and CalOrganics/Greenways) have consolidated ownership over half of the $400 million (USD) in sales the organic produce sector in California generates (Pollan, 2001). Greenways Organic, for example, is a 2000-acre organic produce operation that not only grows organic produce but also controls the packaging stages of production. Greenways products are sold in both the US and Canada. Like Greenways, many organic farms are part of transnational value chains. Earthbound Farm is a producer of bagged fruits and vegetables in North America with over 26 000 acres of agricultural land in California, Arizona and Mexico (Pollan, 2006). Earthbound's business now includes the development of new products for the organic market. By 2003, Earthbound sold over a hundred different items that could be found in over 75 percent of US conventional supermarkets. In 2011, Earthbound began to diversify and launched frozen fruit and vegetable products along with herb purees and snack foods (Earthbound Farm, 2012). Earthbound Farm has vertically integrated (from 'seed to salad') into Natural Selection Foods and contracts 200 growers throughout California (Howard, 2006:18).

While vertical integration is a common corporate strategy to gain control over the organic sector, horizontal integration is an even more popular way of acquiring organic firms. Horizontal integration refers to the acquisition of firms that are involved in similar stages of the value chain and can also include the expansion of a firm's activities (Howard, 2006). It can also be perceived of as a firm expanding within the same stage of the value chain. Heffernan (1998:49) gives the example of the increase in size of farming operations by acquiring smaller farms, which occurred across the US throughout the 1980s. A primary characteristic of this type of integration is that the number of firms reduces, and those firms become larger in capacity. According to the FAO, horizontal integration in the agri-food system is defined by 'a relatively small number of firms effectively controlling a given market' (FAO, 2003). Consolidation of control in one segment of the value chain serves to concentrate profits and can have major implications for competition in markets. Economists use the 'CR4' ratio to determine if levels of ownership concentration are at a point that affects the competitiveness of markets. It is calculated by determining how much of a market share the top four firms control. If it is above 40 percent, the market may be prone to uncompetitive behavior (FAO, 2003; Hendrickson and Heffernan, 2007). The consolidation of ownership of organic firms through horizontal integration by TNCs thus concentrates the profits from sales of organic food without the need to make the costly changes necessary to integrate organic production processes into conventional value chains. Horizontal integration in the organic food sector is present at all stages of the value chain, including the production of organic produce. In 2005, 26 percent of Canadian organic farms earned less than $10 000, while 46 percent were considered 'large', earning over $50 000 (CAD) in 2005. Most of these large organic farms (almost 71 percent) were located in Saskatchewan (Macey, 2004:4). But most of the organic crops grown on the large organic farms in Canada, such as wheat, do not serve domestic markets. Instead, they are bound for the US, the EU or Japan for further processing.

In 2008, 74 percent of the organic foods consumed in Canada originated from the US (ACNielsen, 2009). Most of these products are fresh fruits and vegetables sourced from California. Although California does not represent all state-level organic sectors, it is the source of a significant proportion of organic produce for North America. In 2008, California continued to lead the US states with the largest amount of certified organic cropland in the country (430 000 acres), mainly used for fruit and vegetable production (USDA/Economic Research Service (ERS), 2010). Statistics gathered by the Agricultural Issues Centre at the University of California show that the trend of moving towards larger scales of production of organic products is fully underway in California. In 2000, only one farm grossed $1 million (USD) or above in that year, but that one farm comprised 41 percent of sales of organic products for the state. In 2005, 6 percent of farms grossed over $1 million (USD) and the percentage of gross sales grew to 67 percent (Klonsky and Richter, 2007:11). By 2009, 8 percent of organic growers in California grossed over $1 million (USD) and had 72 percent of total sales (Klonsky and Richter, 2011:10).

A prominent example of the horizontal integration of organic firms at another stage of the production process is Monsanto's acquisition of the organic seed breeder Seminis in 2005. This move by Monsanto further consolidated the seed sector by adding organic seeds to the list of agricultural inputs under Monsanto's control (Organic Monitor, 2003). While Monsanto's involvement in the conventional agri-food sector is formidable, its involvement in the organic sector pales in comparison to the Hain Celestial Food Group, Inc., a conventional agri-food processor that has shifted its business strategy to become the major organic processor in North America. The Hain Celestial Food Group is one of the most aggressive agri-food corporations currently acquiring organic firms through horizontal integration in both Canada and the US.

Before Celestial Seasonings (CS) became part of the Hain Group, it was one of the first independent natural food processors in the US, established in 1970 by Mo Siegel. CS specialized in herbal teas and supplying consumers with 'ecologically sound products' from its Boulder, Colorado base through food cooperatives and health food stores. In the spirit of the libertarianism associated with early alternative agricultural movements, Siegel was intent on keeping unionism out of CS by providing his workers with 'better benefits than any union could' (Belasco, 1989:99–100). By 1978, CS employed over 200 people and made $9 million (USD) in profits. CS became so large that in the late 1970s, it stopped dealing with food cooperatives in the US altogether (Belasco, 1989:99). In that same year, CS expanded its enterprises and developed new overseas supply chains that included production facilities and distributional outlets around the world. It was one of the first US businesses associated with the alternative food movement to go global. Purchased by Dart and Kraft in 1984, CS went on to become worth $10 million (USD), and was still considered by its founder to be a beacon for the alternative food economy. CS was purchased in 2000 by the Hain Food Group that created the Hain Celestial Group (Hain Celestial, 2005).

Before acquiring CS, the Hain Food Group purchased 12 smaller organic processing firms in Canada and the US throughout the 1990s. After acquiring CS, the newly named Hain Celestial Group went on to acquire 16 more organic firms (see Appendix 1). The strategy of horizontally integrating organic firms under the Hain Celestial Group's ownership has proven to be highly profitable. Four years after the Hain Celestial Group joined forces with CS, the Hain Celestial Group collected $544 million (USD) in profits in 2004 (The Hain Celestial Group, 2005; Organic Monitor, 2003). Most of the organic manufacturing facilities purchased by the Hain Celestial Group are located in the US, and most of its organic processing subsidiaries source their raw materials from the US and to a lesser extent Canada (The Hain Celestial Group, 2005:5). The Hain Celestial Group's business strategy is to 'be the leading manufacturer, marketer and seller of natural and organic food … by anticipating and exceeding consumer expectations' (The Hain Celestial Group, 2005:1). This strategy includes further purchasing of organic processors, as expressed by the Hain Celestial Group in its 2004 annual report, declaring its interest in acquiring Spectrum Organic Products, which it did in 2005. Through its major corporate investor, H.J. Heinz (holding 20 percent equity), the Hain Celestial Group has the financial resources to continue its consolidation of the organic processing sector through a strategy of horizontal integration. The Hain Celestial Group now exerts a significant amount of control over the production processes of its organic firms and demands uniformity, standardization and timely delivery of inputs to maintain its competitive edge. In 2008, the Hain Celestial Group announced record sales, up 22 percent from the previous year. It continues to actively acquire firms within the organic/natural food processing segment of the value chain, making it one of the top five food corporations within the organic food sector in Canada and the US. In 2012, the Hain Celestial Group purchased the Daniels Group from the UK, Cully & Sully in Ireland and Europe's Best in Canada (The Hain Celestial Group, 2012:4). Though not exclusively organic food companies, Daniels Group, Cully & Sully and Europe's Best represent significant acquisitions in the global natural food industry as the Hain Celestial Group becomes the largest natural food corporation in the world. Its current tagline is 'organic and natural is in our heart'. The Hain Celestial Group's net sales reached over $1 billion (USD) in 2013, continuing to acquire organic and other natural food companies in the US and abroad (The Hain Celestial Group, 2013:3).

Other major agri-food corporations are actively integrating organic firms into their business activities. US-based Dean Foods, for example, has aggressively acquired the majority of organic milk producers in the US in an effort to secure greater market share, beginning with the purchase of Organic Cow of Vermont in 1999, White Wave (and its Silk Brand soy beverage) in 2002 and Horizon Dairy in 2004 for $200 million (USD). Horizon Dairy and White Wave collectively control 60 percent of the organic milk sector in both Canada and the US (Howard, 2006:18). General Mills, one of the top ten food companies in the US, is another conventional agri-food processor that has acquired numerous organic processing firms and has horizontally integrated them under its management. General Mills' purchases of organic processors Cascadian Farms and Muir Glen has allowed both firms to expand in order to meet rising consumer demand for a variety of organic products. Muir Glen began as a collective farming operation in the 1970s in California. By the 1990s, it had grown into an industrial-scale production facility (Howard, 2009a:17). With the purchase of Cascadian Farms by General Mills, the firm has grown so large that it no longer sources its raw materials from the US, where it was first established. In an effort not to alienate 'alternative' consumers from organic brands owned by conventional corporations, conventional food companies like General Mills have not included any brand identification (the General Mills' 'G' logo) on Muir Glen or Cascadian Farms products. Many parent companies originating from the conventional sector maintain a degree of public distance from newly acquired organic firms. As Howard (2009a:15–18) argues, this is done because of the negative association some organic consumers have of conventional agribusiness. Parent companies do not want to alienate consumers who do not associate the purchase of organic products with participation in an agri-food system that is dominated by agribusiness. This strategy is what Sligh and Christman refer to as 'stealth ownership' (2003:19). Critics of this business strategy applied in the organic sector argue that agri-food corporations' interest in acquiring organic firms is more about improving their public image than changing the way they do business (Cuddeford, 2004; Howard, 2013).

The success of using corporate strategies that integrate pre-existing firms into companies outside of the organic food sector is evident from the rush to secure a market share in organic products in the early 2000s (Table 3.3). Though briefly dipping between 2007 and 2010 due to the global economic recession, the mergers and acquisitions of organic, natural and health food companies continue to steadily increase. By 2008, the top ten food companies in the world had holdings in organic food companies (Table 3.3). In its 2011 Coverage Report on Natural Organic Foods & Beverages, the Capstone Investment Bank reported 21 separate business deals in the organic, natural and health food sector announced in 2010 (Capstone Partners, 2012:5). The more recent trend is for transnational conglomerates to purchase not only natural and organic companies but supplement and vitamin companies. The Capstone report cites that in 2012, 449 mergers and acquisitions of natural, organic or supplement companies took place. The authors attribute the growth in the natural and organic food sector to the perennial trends of growing public awareness of food safety, growing elderly and educated populations and the increased availability of convenience of natural and organic products (Capstone Partners, 2013:2). Corporate efforts to consolidate the natural and organic foods sector are not slowing down and continue to find new ways of integrating organic companies and products into their portfolios.

Table 3.3   *Top ten conventional food manufacturers: investment in*
             organic/natural food companies

Manufacturers and home country	Acquired or hold equity in organic/natural food firms	Date of first acquisition/purchase of equity
Cadbury-Schweppes (UK)	Yes	2002
Coca-Cola (USA)	Yes	2001
ConAgra (USA)	Yes	2000
Danone (FRA)	Yes	2001
Kraft (USA)	Yes	2000
Masterfoods/Mars (USA)	Yes	1997
Nestlé (CH)	Yes	2008
Pepsico (USA)	No*	2003
Tyson (USA)	No*	2001
Unilever (UK/NL)	Yes	2000

* Introduced in-house brand.

Sources: Lang et al., 2006; Howard, 2006, 2009a; Glover, 2005; Draffin, 2006; Organic
Monitor, 2005a; 2005b; Capstone Research, 2012.

Strategic Alliances

Strategic alliances are important ways for agri-food corporations to enter
into cooperative relationships with other firms without purchasing them.
Strategic alliances are established through the coordination of a few firms
to manage stages of value chains, from seed production and genetic
manipulation to the manufacturing, packaging and sale of agri-food
products (Kneen, 1989; Bonanno et al., 1994). Firms that participate in
strategic alliances agree to share resources regarding a particular project
that they mutually benefit from. When a firm participates in a strategic
alliance with another firm, it allows both firms to adjust to new market
conditions more quickly than if they were to enter a market by themselves. For this reason, it has been argued by Sparling and Cook in their
examination of strategic alliances between agri-food companies under
NAFTA that strategic alliances are less financially risky than vertical or
horizontal integration (Sparling and Cook, 2000:91). William Heffernan,
in his examination of corporate activity in the industrialized agri-food
system, identifies strategic alliances or *network clusters* that are used to
reduce market competition through cooperative behavior between groups

of firms (Heffernan et al., 1999:3). Network clustering emerged in the 1980s and refers to the oligopolistic concentration of ownership among a few corporations and their cooperation with each other to gain and protect market shares. Establishing strategic alliances as a competitive market strategy was part of the reason that market share of the top 20 US food manufacturers doubled between 1967 and 2002 (Lang, 2003:18).

Strategic alliances differ from vertical integration, as they are relationships between corporations based on the coordination of production processes and not the centralization of control over all production processes. Cargill and IBP's relationship in the early 2000s is an example of agribusinesses establishing cooperative relationships with one another to control production processes while sharing in the profits. Cargill and IBP together controlled 74 percent of Canadian beef packing plants and the vast majority of beef packing, corn exports, soybean crushing, soybean exports, flour milling and pork packers in the US (Howard, 2006:17). This trend of concentration of ownership in the beef packing industry continues in both Canada and the US (Rude, Harrison and Carlberg, 2010). US-based agribusiness Archer Daniels Midland (ADM), for example, made a significant effort to incorporate Canadian processors into its portfolio and attained control over 30 percent of Canadian flour milling capacity in 1995, which expanded to 40 percent in 2003 (Qualman and Wiebe, 2003:12).

One of the best examples of a strategic alliance in the conventional agri-food sector is the alliance between Monsanto and Cargill (Heffernan et al., 1999:6). Instead of participating in direct competition with one another, Monsanto and Cargill agreed to cooperate by exchanging information while concentrating in separate stages of the supply chain. For example, Cargill and Monsanto have formed a cluster in which Monsanto provides genetic material and seed while Cargill performs the grain collection and processing (Howard, 2006:18). Monsanto's strategic alliance with Cargill has extended to the organic food sector with the establishment of Cargill's strategic alliance with French Meadow (a US-based organic bakery) in 2002 and Hain Celestial in 2003 (Howard, 2013). Strategic alliances enable conventional agri-food corporations to share control over various stages of the organic supply chain without the same financial risks as integrating organic firms through acquisitions.

Brand Introduction

Another strategy used by corporations to enter the organic sector and boost sales is the diversification of product lines. To meet rising consumer demand for more healthy foods while maintaining consumer loyalties to particular brands, corporations in the late 1990s began to make improvements on pre-existing product lines (calorie reduction, addition of nutrients, fat free) and introduced new brands of agri-food products with organic qualities (Lang and Heasman, 2005; Nestle, 2003). Howard has labeled these strategies 'concentric diversification' (Howard, 2009a:21). Instead of integrating vertically or horizontally, or aligning strategically with other firms, conventional agri-food companies use organic branding and labeling to capture market shares. In the food sector there are two ways a product brand can be introduced: through a food manufacturer or food retailer. A manufacturer's brand is the product of the manufacturer and can be distributed throughout retail outlets. A retailer-introduced brand is owned by the retailer and is only distributed through that retailer. Like other firms that are part of the agri-food value chain, food retailers desire to have greater control over pricing to compete with other food retailers (Coleman et al., 2004:42). Most food retailers carry a mix of manufacturer and retailer-introduced brands of food products (Burt and Sparks, 2002).

Food manufacturers introduce new brands for a number of reasons including garnering consumer loyalty, achieving wider distribution and differentiating their products from the competitors (Lang and Heasman, 2005:156). Consumers may associate quality with certain food labels, so introducing a new brand may capitalize on pre-existing consumer loyalty. Successful brands gain consumer loyalty because they deliver added value beyond merely meeting the criteria of what constitutes a particular product (Burt, 2000). Once consumer loyalty is established, food manufacturers can build on this loyalty by increasing the prices of the brand-name products since consumers associate the brand with a certain level of quality and taste that the competition does not possess. Consumer loyalty to a particular brand can also increase the distribution of manufacturers' brands because of consumer demand. Corporations embark on the strategy of brand introduction for the purposes of product differentiation, and building on consumer loyalty to particular brands offers consumers other varieties of products using the same label. Notable examples include the introduction of Ragu Organic in 2005, Ben and Jerry's Organic in 2003, Campbell Soup's introduction of Campbell's Organic in 2003 and ConAgra's introduction of both Hunt's Organic and Orville Redenbacher's Organic in 2005 (Howard, 2013). By introducing an organic version under the same brand name, consumers can identify with a familiar label, like Ragu or Campbell's, but also purchase a food product with additional characteristics that are perceived as healthier.

In addition to agri-food manufacturers and processors introducing organic brands, many major food retailers introduce in-house brands that carry their name. Retailers that introduce in-house brands can capitalize on consumer loyalty and consumers' association of the retail outlet with a certain level of quality. For example, Canadian supermarket chain Loblaws introduced its President's Choice Organics line in 2001. In a similar fashion Canadian and US food retailer Safeway introduced its own brand called 'O Organics' in 2005 (Howard, 2006:18). Since both Loblaws and Safeway carry organic in-house brands and manufacturer brands that share similar material characteristics (both use certified organic ingredients), price may be the ultimate determining factor for a retail food shopper who wants to buy organic products but is otherwise discouraged by the premium price (Thomson, 1998; Walnut Acres, 2005). Both retailer brands were introduced with the objective of offering, at lower prices, a variety of organic products that were previously avoided by many of their customers because of the premium price (Weeks, 2006: D4). Other conventional retailers in Canada have begun to sell organic food, including Thrifty Foods, Sobeys, A&P and Metro. Sobeys introduced its own in-house organic brand in 2006. In 2008, private in-house labels accounted for 21 percent of sales of organic products in Canada (Agriculture and Agri-Food Canada (AAFC), 2009). As with conventional agri-food processors, the top ten global food retailers including Tesco, Ahold, Carrefour and Metro have introduced organic food products onto their shelves to benefit from organic food's premium prices (Glover, 2005; Draffin, 2006; Lang et al., 2006). In the US, large-scale food retailers like Kroger, Albertson's, Costco and Wal-Mart are now embracing organic products and are increasing the number of products they carry, although none of these food retailers have introduced their own in-house organic product lines.

The increase of food retailers' incorporation of organic products into their business strategies has resulted in a shift of organic sales from the small-scale outlets to large-scale corporate retailers. The integration of organic products into conventional food-retailers' stock has effectively changed where people purchase organic products. This shift in purchasing habits is reflected in the growth of sales of organic through large-scale food retailers over the last decade. According the US statistics, in 1991, 7 percent of all organic products were sold in conventional supermarkets and 68 percent were sold in health food/natural products stores. As a result of the increased investment of agri-food companies in 2000, 49 percent of all organic food products were sold in conventional supermarkets and 48 percent were sold in natural food/health food stores (with 3 percent through direct to consumer methods, for example, box schemes) (Dimitri and Greene, 2002:2).

Although similar longitudinal statistics are not available for Canada, sales of organic products in 2006 exhibit similar trends observed in the US in 2000. According to Macey's 2007 report to the Organic Agriculture Centre of Canada, 41 percent of all sales of organic products in Canada were through conventional food retailers. In 2008, this number increased to 45 percent, amounting to $925.8 million (CAD); 33 percent were purchased through natural food retailers (Macey, 2007:2). Direct consumer sales increased to 20 percent of the market share in 2008, which was worth $400 million (CAD). In 2005, the Hain Celestial Group reported that $45 billion (USD) of its sales of organic and natural foods was through 60 000 conventional food retailers across Canada and the US, making conventional food retailers increasingly important actors in organic value chains as organic food gains a larger share of the retail food market (The Hain Celestial Group, 2005:4). By 2012, supermarket sales of organic packaged foods rose to 72 percent share of market sales in the US as demand had increased for these types of organic food products (Euromonitor, 2013:9).

Corporate Success in the Organic Food Sector: Whole Foods Market

The merger and acquisition strategies of conventional agri-food corporations in the organic sector have had significant impacts on what defines organic, both as a product and a social movement. It is important to note, however, that involvement of conventional corporations has also occurred within the organic sector itself, revealing the broad impact of the economic success and expansion of organic, and the pressure imposed on the viability of the process-based definition of organic. Whole Foods Market (WFM) is in many ways the archetypal business for defining the corporatization of the organic food sector. WFM first emerged in 1980 with the merger of Safer Way Natural Foods and Clarksville National Grocer in Austin, Texas. It began as a small natural and organic food retailer that sourced most of its products locally. WFM is a self-described 'natural food retailer' and promotes itself as the alternative to conventional food retailers. WFM posts photographs and biographies of organic producers in its stores who supply the company with its vegetables and fruit, informing consumers of the origins of the organic products (Kabel, 2006). The primary goal of WFM in its own words is to 'make grocery shopping fun', to make food purchasing an experience as opposed to a necessary chore (WFM, 2005b:1). Its success as the leading 'natural' and 'organic' food retailer has also made it the most profitable natural foods retailer in North America, selling $12.9 billion (USD) of organic and natural foods in 2013 (WFM, 2013).

WFM's promotion of itself as an alternative food shopping experience has paid off and attracted millions of consumers in Canada and the US, contributing to its rapid expansion and growth (Burros, 2007). In 1991, there were ten WFM stores in the US. By 2005, there were 175 and this number continues to expand. In 2005 WFM was the twenty-first largest supermarket (by sales) in Canada and the US, and ranked 479th of all US companies based on sales (Howard, 2006:18; WFM, 2005b:4). Between 1991 and 2007 WFM acquired 14 other organic/natural food retailers in the US, and over the last 27 years WFM has purchased 18 other smaller companies in attempts to consolidate ownership over the natural/organic foods retailing sector (Associated Press, 2007).

WFM's successful business strategy has fostered its growth as the biggest natural and organic food retailer but the desire to consolidate the distributional segment of organic value chains under centralized management means that the company exhibits oligopolistic behavior similar to that of corporations found in the conventional agri-food sector. In addition to expanding its operations across Canada and the US, WFM is also expanding its presence on the other side of the Atlantic, purchasing UK-based Fresh and Wild supermarkets in 2004. Early in 2007 WFM proposed a merger with Wild Oats Markets, its major rival in the US (WFM, 2005a; Martin, 2007). In June 2007, the US Federal Trade Commission (FTC) granted a temporary injunction to block the merger bid from WFM for Wild Oats (Associated Press, 2007). The concern by the FTC was that the merger of WFM and Wild Oats would result in a monopoly of the natural foods distributors. Eventually, the case went to district court and the judge turned down the FTC's move to block the merger arguing that the merger did not violate any US anti-trust laws. WFMs went ahead and purchased Wild Oats Markets for $565 million (US) in August of 2007. By the end of 2007, WFM was operating 276 stores across the US, Canada and the UK (A.W. Page Society, 2009). By early 2015, WFM had 386 stores across the US, ten in Canada and nine in the UK with over $12 billion (USD) in retail sales (WFM, 2015; Van Praet, 2013).

WFM's internal expansion demands that its suppliers be able to handle larger purchase orders and provide standardized, dependable quality goods on a large scale in order to take advantage of economies of scale (Mark, 2004). In addition to practicing business strategies used by conventional food retailers like Walmart, WFM has taken similar steps to keep labor organization out of its business model. In its earlier years, WFM practiced what has been described as 'democratic capitalism', using self-directed teams of employees and creating what is described as a horizontal form of labor organization (Mark, 2004; Dimitri and

Richman, 2000:14). Much like Walmart's 'associates', WFM does not call its workers 'employees' but 'team members'. New employees are given a four-month trial period before their 'team members' vote on whether they should remain on the team. Yet, as it expanded and became more profitable, WFM's employees were not satisfied with WFM's model of 'democratic capitalism' and became interested in unionizing. But unfortunately for WFM associates, John Mackey, the co-CEO of WFM, shares a similar attitude towards unions as the first CEO of Celestial Seasonings, Mo Siegel. Mackey considers unions 'parasites' and has responded to criticisms of WFM's business strategy by noting that his company is in the business of whole foods not 'holy foods' (Harris, 2006:62); he is quoted (in Guthman, 2004:110) as asking, 'where in our mission statement do we talk about trying to be liberal, progressive or universal?'

Despite being listed as one of *Fortune Magazine*'s '100 Best Companies to Work for in the US', WFM has been embroiled in a number of labor disputes (Sharpin, 2006). Workers at WFM are not part of a trade union and, according to the website 'wholefoodsworkersunite.org', WFM is quite hostile toward unionization. A Madison, Wisconsin store attempted to unionize but failed because the WFM's executive management blocked the effort. One member's contribution to the website (wholeworkersunite.org) states that:

> many of us have ... seen that as the company has grown, the focus has shifted to profits and expansion at the expense of worker respect and fair compensation ... Despite what WFM says, unionizing is the only way for workers to be guaranteed participation in their employment.

In response to workers' complaints regarding questionable labor practices at WFM, the United States Department of Labor (USDL) took WFM to court over $226 000 (USD) in overtime wages that had not been paid to some of its 'team members' (FamilyFarmDefenders, 2007). The rejection of unionization and other efforts to assure fairer labor practices within WFM draws attention to the diverse range of commitments to social sustainability within the organic sector. Despite the fact that there is a long history of organic practitioners embracing free market principles, others holding the process-based interpretation of organic see social responsibility as an integral part of what makes a good organic. What is considered and included in a vision of social responsibility is not easily identifiable in the contemporary organic food sector. A commitment to social responsibility is by no means a coherent position across the sector.

IMPLICATIONS OF THE CORPORATE MODEL IN ORGANICS

Supporters of the incorporation of organic food into the conventional agri-food system argue that when conventional agri-food corporations purchase organic firms, organic products get wider distribution and give consumers cheaper access to healthier food (Klonsky, 2000:242). Consolidating production processes helps to streamline processes that may contribute to lower production costs in the long run. From the perspective of maintaining a connection between productive processes and the process-based definition of organic, however, the involvement and competition of conventional agri-food corporations in the organic value chain presents a number of challenges (Buck et al., 1997; Howard, 2009b). As Clunies-Ross notes, 'paradoxically, just as consumers are beginning to make a negative link between food quality and the industrialization of the food process, attempts are being made to draw producers of organic food into the commercial food sector in an effort to meet consumer demand' (Clunies-Ross, 1990:212). The corporatization of the organic sector has a number of implications for the viability of the process-based concept of organic, specifically the commitments to disaggregated decision-making, environmental considerations and social sustainability.

Decision-making

Supporters and practitioners of the process-based definition of organic embrace an alternative form of market interactions and activities. One of the original goals of organic agriculture was to keep ownership of organic businesses disaggregated to allow for a large group of supporters of organic values to work together to foster a widespread, sustainable agri-food system. A review of Rodale's *How to Grow Vegetables and Fruits by the Organic Method* in the last issue of the *Whole Earth Catalog* (1971) demonstrates an early example of how practitioners challenged the status quo by promoting a 'do-it-yourself' mantra as a fundamental component of the organic movement. At the end of the piece, reviewer Gurney Norman comments on the important social role that Rodale's periodical *Organic Gardening* has on the movement as a whole. Norman (1971:50) states:

> [I]f I were a dictator determined to control the national press *Organic Gardening* would be the first publication I'd squash. I believe the organic gardeners are in the forefront of a serious effort to save the work by changing man's orientation to it, to move away from collective, centrist, super-industrial

state, toward a simpler, realer one-to-one relation with the earth itself ... The thing I like to remember is that even when all the froth has blown away, and the rhetoric of pop-ecology has drifted off to join the other forms of pollution in the sky, the gardeners are going to still be gardening. They're going to quietly go on composting and tilling and planting, and then reaping all the good things they have sown.

Practitioners of organic techniques in the early 1970s associated organic food production with a form of social independence from and social resistance to the conventional economic and political system. Supporters were encouraged to view organic agriculture as a form of resistance to the industrial model of food production's consolidation of control and resources within the agri-food system.

Evidence of the democratic spirit can also be found in early charters from organic agricultural associations like California Certified Organic Farmers (CCOF), discussed in more detail in Chapter 4. The CCOF would go on to be an example for other organic farming associations in the US and Canada in the 1970s. The original principles listed in the first issue were developed by farmers, for farmers. For example, principles 4, 5 and 6 of the CCOF's first published standards (CCOF, 1974:16) expressed that each member of the CCOF was given a significant degree of responsibility to protect the CCOF official seal, as individual members were responsible for following the rules without intrusive overhead surveillance from CCOF:

4. Each member will market or sell his produce by the best method suitable to maintain freshness, quality and appearance.
5. All organically grown food marketed by a C.C.O.F. member must be identified by the official C.C.O.F. seal.
6. A member may sell any food or produce he raises not meeting these standards... but, the member is responsible for protecting the C.C.O.F. name by making certain the buyer is not left with the impression that he is buying food certified by C.C.O.F. ... If a member is found to be in violation of this section, he is subject to immediate suspension, suspension of his seal, and ultimately expulsion from this organization upon majority vote by the appropriate committees.

Once a member agreed to follow the standards, it was up to him/her to uphold them. Though inspections by the CCOF did occur, if a member was found to be noncompliant there were ramifications as outlined in Principle 6. For some early practitioners, organic agriculture was a socially conscious action that required inclusivity and responsiveness to the needs of the people involved in the production process. Decision-making over acceptable practices for organic practitioners was disaggregated and heterogeneous among many people since broad consensus is a fundamental element to organic culture. In many cases, pioneering organic firms put into practice these values whether they were cooperative retailers or collective farming operations. With the consolidation of decision-making power by publicly traded corporations that are primarily responsible to shareholders expecting growth in their investments, organic firms are oriented towards profit maximization and the reduction of overhead costs. As Howard notes in his contribution to the publication *Natural Farmer*, the consolidation of power over the agri-food sector by a small number of firms gives them 'disproportionate influence on not just price, but also the quantity, quality and location of production' (Howard, 2006:18). Concentrated ownership of organic firms has thus meant a loss of independence for organic producers in determining how production processes should be organized, and what they should include (Welling, 1999:41). Corporatization of agriculture threatens the ability for individual small-scale farms to co-exist with corporate actors; much like in the conventional agriculture sector in the 1980s, many organic producers are now forced to either 'get big, or get out' in order to stay competitive.

The consolidation of power over the agri-food system contributes to what is described as 'food from nowhere', as the majority of the decisions about what food is produced and how it is produced are usually made far away from the points of production, processing and distribution (Bove and Dufour, 2001:55). Transnationalization of value chains thus erodes the role of trust in the organic sector that is built through localized agri-food networks where organic practitioners are often in direct contact with consumers. As a result, standards, coding systems, means of regulatory enforcement and penalties for violations are increasingly necessary for sustaining the consumer confidence (Allen and Kovach, 2000:223). Locating various aspects of the production process around the world encourages further formal regulation to allow firms to participate in organic agriculture because the local, personal relationships that characterize more traditional forms of organic agriculture are impossible to sustain.

Environmental Considerations

Practitioners who adhere to the product-based definition of organic claim that despite not accounting for environmental costs throughout the production process, the reduced amount of synthetic inputs used in all types of organic farming aids in reducing dependence upon non-renewable resources and environmental degradation. Lynch's 2009 review of studies focusing on the environmental impact of organic agriculture in Canada and the US found that crop yields are an ongoing challenge but that organic systems positively contribute to ecological and environmental goals (for example, diversity of crops, farm nutrient intensity, reduction in energy, reduction in pesticide use) (Lynch, 2009:626). Studies comparing environmental impacts of organic agriculture compared to conventional agriculture demonstrate that, to some degree, Lynch's claim in the European context is true. Tuomisto et al.'s (2012) analysis of 71 reports on the environmental impacts of farming (comparing nutrient losses, biodiversity impacts, greenhouse gas emissions, eutrophication potential, acidification potential, energy use and land use) shows that organic practices are environmentally beneficial on a land per-unit basis. However, these practices have less of a positive impact on a product per-unit basis. Organic practices tend to add organic matter to the soil per-unit of field area, but emissions such as ammonia, nitrogen leaching and nitrous oxide product per-unit were higher. Organic systems also had lower energy requirements compared to conventional land requirements. The general decline of biodiversity is argued to be a result of the scale of farming operations as opposed to the type of farming performed. Overall, the authors state that the environmental benefits of organic agriculture is highly dependent on what type of crop is being produced, making it difficult to come to any general conclusions about the environmental sustainability of large-scale organic agriculture compared to small-scale organic agriculture (Tuomisto et al., 2012:318).

Holistically, organic agriculture as a production system has a lower environmental impact. However, these gains are undermined by the de-emphasis of the product-based approach to environmental goods throughout the production process. When evaluated on a per-product basis (excluding process elements such as biodiversity), the benefits compared with conventionally produced products decline, especially if organic is practiced on the same scale as conventional. To meet growing consumer demand, the approach to organic production processes that emphasizes the qualities of the product over the value of the process does not address the environmental costs associated with industrial technologies dependent on fossil fuels and other structures of the conventional agri-food system such as transnational value chains and monoculture. Practicing monoculture on a large scale is consistent with the product-based approach to organic production because, unlike using GMOs and synthetic inputs, it does not impact the material qualities of

the end organic product and is not a consideration for organic certification. Ronnie Cummins of the OCA expresses concern at the lack of
emphasis placed by practitioners of the product-based definition of
organic on the environmental implications of large-scale organic agriculture. As Cummins charges:

> no way in hell can you be organic if you have over a few hundred cows. After
> a certain size, the operation cannot be ecologically sound anymore ... large
> monocultures, using large energy inputs and receiving subsidized water ...
> [are] three elements that are anti-environmental and unacceptable for those
> who want ecologically sound farming. (quoted in Ruiz-Marrero, 2004)

Through the practice of monoculture, the environmental sensitivity that
supporters of the process-based definition of organic put into practice are
hollowed out by large-scale organic monoculture that does not consider
biodiversity and mixed farming as integral to the definition of organic.

The desire to widen the distribution of organic food as prescribed by
the product-based approach to organic increases the 'food miles' traveled
by organic products. Tim Lang first introduced the concept of 'food
miles' in 1994. According to Lang, food miles refer to the distance that
food travels from the farm to the plate and includes all of the steps in
between. Those critical of the global transportation system for food claim
that every calorie of food energy produced in the conventional system
requires ten more calories of energy to transport it to its destination
(Imhoff, 1996:426; Lang and Heasman, 2005:235–7). In 1994, the
Sustainable Agriculture Food and Environment (SAFE) Alliance published 'The Food Miles Report', documenting how the globalization of
the agri-food system was causing environmental pollution through the
use of conventional transportation networks (Paxton, 1994). This report
helped to launch the idea of local food chains as a challenge to
globalized food chains; the fewer food miles traveled, the more 'green'
and environmentally conscious is the food. The concept of localized food
chains was linked with organic and promoted as the most ethical way to
eat. To some degree, there is validity to linking the criticisms of food
miles with the organic movement's support of localized agri-food chains.
Critically examining the linkages between socio-economic and environmental factors related to the distance between producers and consumers
as well as food transportation networks is vital to discussions involving
social justice and progressive politics within the food system. In terms of
the process-based definition of organic, by denying the importance of
where an organic product is produced, the product-based approach to
organic production diminishes the reduction of the environmental impacts

of the agri-food system as a primary goal of organic agriculture (Powell, 1995:122). With its insertion into global transportation networks, many of the environmental goods associated with the process-based definition are lost since non-renewable resources fuel the extensive transportation network necessary for a globalized agri-food system (Bentley and Barker, 2005). By using the same global transportation networks as conventional agri-food, one contributor to the Organisation for Economic Co-operation and Development (OECD)'s report on organic agriculture notes, 'the environmental credentials of organic products are compromised where they are transported over long distances' (Hallam, 2003:186).

Yet, the argument that localized food chains are more environmentally friendly than global food chains is not without its critics. In 2005, a report from the Department of Environment, Food and Rural Affairs (DEFRA) (UK) entitled 'The Validity of Food Miles as an Indicator of Sustainable Development' challenged the idea that globalized transportation networks were more environmentally damaging than localized food chains. The research team considered car trips to the supermarket may create more pollution than by the air travel required to get the food to market (Smith et al., 2005). More recent studies claim that transportation networks of chain supermarkets may be more efficient in the use of fossil fuels necessary to get food products to consumers than the energy necessary to get a pound (in weight) of food to a farmer's market (Nicholson et al., 2011). Whether locally sourced and distributed foods are in fact more environmentally friendly is an ongoing discussion. Supporters of the process-based concept of organic maintain that local agri-food systems have the ability to produce food that avoids many of the environmental hazards created by globalized transportation networks.

Labor Practices

The transnationalization of value chains problematizes organic food's commitment to social sustainability by relocating various parts of production processes to wherever the lowest labor costs may be found (Bonanno et al., 1994), or substituting local labor with foreign-domestic labor pools that often cost employers significantly less to employ. In essence, products produced abroad may not necessarily meet the same standards as those applied domestically, compromising the commitment to the substantive social aspects of the process-based definition of organic and the multifunctional integrity of the certified organic label in a globally integrated value chain.

Certification standards for organic agriculture are primarily focused on material inputs, and in the case of Canada, to some degree animal welfare, not labor inputs. Labor practices on the farm or along the value chain are not considered in certification evaluation. Organic practitioners are not necessarily socially progressive and do not always adhere to principles of social responsibility on their farms or in their business practices. Since certification does not dictate any labor requirements or restrictions for a good to be considered organic, corporate actors, or those who do not support progressive labor practices through intervention in market activities, can participate in the organic agri-food sector without addressing labor conditions within the value chain. A study from the University of California shows that workers on organic farms are not necessarily better off than those working on conventional farms as common understandings of organic agriculture suggest (with one major exception being exposure to agricultural chemicals) (Mark, 2006; Blum, 2006; Getz, Brown and Shreck, 2008). In the case of Southern California, agricultural laborers in the organic sector are sometimes exposed to dangerous, unsanitary conditions for wages that often do not meet state minimum wage legislation or state regulations pertaining to hand weeding. Similarly, in Arizona Roane reported questionable labor practices in the organic sector, including worker complaints of being threatened by employers, unfit living conditions and children less than 14 years of age working on organic farms (Roane, 2002).

Corporate consolidation has spillover effects for the treatment of labor on organic farms operating on a smaller scale as well. Shreck et al.'s 2006 University of California study is based on interviews with organic producers in California and shows that small-scale organic producers claim they are unable to provide benefits for their workers because they do not earn enough profit to provide benefits for themselves, let alone their workforce. Thus, the prices that organic farmers receive for their products are not enough for farmers to be 'socially sustainable' with regard to their labor force as corporate competitors that can meet economies of scale offer more competitively priced organic products (Shreck et al., 2006; Daily Democrat, 2005). Critics of this position claim that 'farm-owners may argue that they cannot afford to provide better conditions where labor costs increase and opportunities ... but additional costs may be more than offset by savings in other external inputs and higher prices' (Blowfield, 2001:5). Nevertheless, the influence of corporate consolidation and presence of cheaper organic products has overarching structural implications for social sustainability, both on corporate farms as well as organic foods produced on a smaller scale.

CONCLUSION

Though the methods of organic agriculture have influenced corporate activities to some degree, the ability to put the process-based definition of organic into practice has been seriously compromised as firms within the organic sector continue to aggressively consolidate ownership by employing conventional corporate strategies well into the 2010s. The practices of firms from within the organic sector are also influenced by corporate strategies outside of the sector, such as WFM. The co-existence of two approaches to organic within the organic food sector presents challenges to those committed to substantive organic values as they compete against organic firms focusing on the qualities of the organic product that can often produce similar goods on a larger scale and a reduced cost. It creates consumer confusion, as the organic label is assumed by some to indicate that the product is produced in adherence to traditional organic principles, when in reality it may not be the case. As Vos notes, '[the] ideological lineaments of organic farming … represent an historically persistent cultural paradigm … [yet] … this paradigm may be increasingly called into question' (2000:252). The insertion of the organic sector into the conventional market economy imposes significant contradictions into the institutional development of the organic sector, including organic agriculture itself.

Corporate consolidation of ownership in the organic sector has occurred at a number of stages of the value chain, including seed distribution, processing and the distribution of agri-food products to consumers. Acquiring organic firms, forging strategic alliances and introducing 'in-house' organic brands are effective strategies for conventional food corporations to concentrate ownership and control over organic production processes. In today's globalized economy it is unrealistic to attempt to re-localize all organic value chains or insist that if one has the desire to consume organic products then she or he should only purchase those grown or processed locally or by small-scale businesses. In some cases, this is simply not possible. The globalized agri-food system has brought food choices to people around the world who can afford variety in the food they purchase and eat. But it is highly problematic for conventional agri-food corporations to continue to draw upon the mythology of the 'local' and the 'sustainable' small-scale organic farmer to promote and sell premium-priced organic products, when local and sustainable may be furthest from the truth.

4. From private to public: institutionalizing organic food standards into policy

> Any farmers group that gets together and verifies itself is unacceptable.
>
> Tom Harding, president,
> Organic Food Production Association
> (quoted in Burros, 1987)

INTRODUCTION

In response to increased production and consumption of organic foods, both Canadian and American governments embarked on formalizing organic agricultural standards into national policy in the late 1990s. Organic agriculture in both contexts came to be viewed by national governments as an economically valuable agricultural sub-sector that required public regulation and funding to expand (Haumann, 2010:183). To replace the patchwork of private certifiers, as well as various provincial and state regulations and standards, the US government passed legislation creating a national organic program in 2000, while Canada formally passed its national organic standards in 2009. Many actors welcomed the development of national organic standards as they are argued to give consumers a reliable, recognizable label by giving producers a standardized set of guidelines and regulations.

Although regulating the production processes involved in organic agriculture is itself not new, the institutionalization of standards and principles into public policy in Canada and the US is a very recent development. What sets the institutionalization of rules surrounding organic food and agriculture in national public policy apart from prior forms of regulation is the ascendancy of the idea that organic agriculture should be regulated through a series of enforceable legal frameworks that, for the most part, do not include some of the substantive issues included in the process-based definition of organic, such as labor conditions or farm size.

This chapter examines how, as organic production processes began to be regulated in multiple levels of public policy in Canada and the US, the substantive goals of the process-based definition of organic were excluded from the institutionalized definition of 'organic', giving priority to the product-based definition. The first section discusses factors contributing to the shift of organic regulations from the hands of private, grassroots organic producers' organizations to state and provincial governments. It traces the trajectories of both conventional and organic agri-food policies to demonstrate that they initially developed in different contexts with distinct and often incompatible goals and systems of decision-making. The second section examines the continued expansion of the organic regulatory regime to the national levels of policy making in the US and Canada. The development of national regulations for organic agri-food was primarily oriented towards facilitating the expansion of international markets to export domestically produced organic products and to standardize a set of enforceable, regulated production principles behind national labels. As the rules and principles regarding organic agri-food have become institutionalized into federal policy frameworks, the regulations for organic agriculture have privileged harmonized trade, efficiency and market-competitiveness over the social and environmental goods associated with the process-based definition of organic.

THE MOVE TOWARDS REGULATION

Three Transitions in Organic Agriculture Policy

There have been three major transitions in the development of organic agriculture policies since practitioners of organic agriculture first began to come together in the 1960s to create organic agriculture organizations. Initially, organic producers established informal groups based on their shared principles outlining how organic agriculture should ideally be practiced. Local communities of organic practitioners organized groups to produce their own sets of guidelines. Adherence to these guidelines was maintained predominantly through trust-based relationships. The second transition occurred as markets for organic food expanded and distances between producers and consumers grew. More formal private associations emerged to formulate producer 'codes of conduct' and established production standards. These first two transitions were largely led by member-based organizations consisting of those supportive of putting the process-based definition of organic into practice. The third transition was the integration of private sector organic regulations into public policy frameworks, resulting from growing consumer interest in organic products and the entry of new actors into the organic sector in the late 1970s. The main purpose of public regulation since it emerged in the organic sectors in Canada and the US has been to monitor the authenticity of organic products to assure consumer confidence while facilitating the expansion of markets and satisfying trading partners' (such as the European Union) demand for nation-level standards.

The Industrialized Model of Food Production: The Rise of 'Cheap Food' Policies

The emergence of informal grassroots organizations of organic producers was a response to the privileging of industrialized agriculture by federal governments in the post-war period. During this time, making food affordable and accessible were of prime importance to the American and Canadian governments as food shortages were experienced during World War II. In 1933, for instance, Americans used more than 25 percent of their disposable income on purchasing food. Since 1947, this number has steadily declined (Miller and Coble, 2005). At the time, governments reasoned that the food supply of domestic populations was too important to leave up to unbridled market forces (Giangrande, 1985; Skogstad, 1987). Agricultural policies were but one pillar of Keynesianism that flourished in the post-war period until the mid-1970s that supported the intervention of government into market activities.

Under Keynesian economic strategies applied to the agricultural sector, national governments control supply or provide price supports to stabilize prices for farmers as well as for consumers. Keynesian agricultural policies also manage the supply of agricultural goods destined for the export market. An issue related to Keynesian agricultural policies is labor market policy, which under this system prioritizes programs and strategies that keep national unemployment levels low and provide income supports when citizens experience involuntary unemployment. Keynesian agricultural and labor market policies worked together to stabilize prices for farmers while supporting employment in the agricultural sector. The Keynesian approach to managing agricultural sectors allowed enhanced efficiency and industrialization of agricultural techniques to coexist with relatively diverse and small-scale patterns of ownership. Efforts to secure a system that produced a plentiful and reliable food supply were successful partially due to the wide-scale adoption of industrial agriculture technologies, but also because industrialized methods of food production were favored and financially supported by agricultural departments. The cost of food for consumers in Canada and the US declined steadily over the next several decades.

Privileging industrialized agriculture in government policy, however, did not go unchallenged. Critics pejoratively refer to the system of rules governing agricultural management in North America as 'cheap food policies' (Mitchell, 1975). This term refers to the state's effort to support increased agricultural outputs while lowering food prices for consumers through industrialized methods of agricultural production. It is defined by Knutson et al. (1998) as a system 'that involves the government overtly pursuing policies that hold down the price of food below the competitive equilibrium price' (quoted in Miller and Koble, 2005:2). Cheap food policies have made it possible for citizens in both Canada and the US to drastically lower the overall percentage of income they spend on food. The over-production of commodities facilitated by supply side management policies helped to keep some commodity prices low and forced farmers to produce increasingly higher volumes in order to earn enough income to support themselves. Cheap food policies encouraged the rapid industrialization of North American agricultural systems in order to produce increasing amounts of food (Clunies-Ross, 1990). There were no formal limitations on practicing non-chemical farming, but it made little economic sense for most farmers because government support was contingent on high farm yields and uniform outputs, which were only attainable through industrialized organization of production.

As cheap food policies helped to develop an industrialized agricultural system and keep food prices low for consumers, criticisms of the agri-food system began to emerge in the 1960s. Rachel Carson's *Silent Spring* (1962) was one of the first publications to sound the alarm regarding the environmentally destructive outcomes of industrialized agriculture and particularly the use of DDT in farming. Having worked as a marine biologist for the US Department of Interiors for 17 years, Carson took notice of the changing ecology around agricultural stretches of land in the US. Her criticism of industrial agriculture's contribution to the declining presence of songbirds across the US was a poignant moment for the environmental movement. Her book drew wider public attention to the environmental damage that industrialized agriculture caused to the American landscape. The attention paid towards the environmental consequences of industrial agriculture sparked wider public discussions about how human activities were causing environmental damage that was affecting the health of humans, plants and animals.

Negative social consequences of industrialized agriculture were also gaining public attention. These came in the form of rural unemployment, farmer indebtedness and a loss of control held by individual farmers over what they produced and how they produced it (Giangrande, 1985). Although supply-side agricultural management policies in the early stages of their development did address some social issues such as shielding farmers from fluctuating commodity prices (and therefore providing them with a relatively stable income) and unforeseen 'forces of nature' through crop insurance, other elements of the production process (particularly the ecological problems created by the industrialized agri-food system) were overlooked by governments in their quest to secure a stable, cheap food supply (Berry, 1977; Warnock, 1987).

The early organic movement (largely a product of other 'new' social movements emerging in 1960s including the Environmental and Back to the Land movements) was a critical reaction to the social and environmental consequences of the industrialized system of agriculture (see Chapter 6). Because of the corporatist links between industrialized agriculture and government policies subsidizing it, organic practitioners sought political and economic autonomy from the state (Guthman, 2004). Independence from state intervention and financial programs allowed private actors to develop norms of organic agriculture that were responsive to their needs and objectives, thereby allowing them to practice a type of food production that internalized and sought to reduce a number of the costs associated with agricultural production. The desire for distance appeared to be mutual between agricultural departments and organic producers who wanted private sector rules for organic agriculture to develop independently from the state (Merrill, 1976). At this time, agricultural departments in Canada and the US did not consider organic agriculture as a type of food production that should be supported or promoted as a viable alternative to industrial production.

Early Organic Producers Associations

Early regulations for organic agriculture were formulated in sometimes overlapping stages and included the principles of diversity and a commitment to horizontal decision-making structures, both reflecting some of the principles found in the process-based definition of organic. Most organic agriculture in the early post-war period was geared towards self-sufficient food production as organic producers insisted on keeping a distance from government subsidies and agribusiness, restricting any form of control that either could have on production standards and practices. Instead, organic producers used the resources available to them (physical labor, compost, social networks and agronomy texts) to nurture the health of the soil. From the desire to remain independent and in control of defining what constitutes 'organic agriculture', informal grass-roots organizations established their own set of guidelines, standards and practices.

Although more formal organizations developed in the late 1960s, the first major effort to organize practitioners of organic agriculture occurred at the international level. Lady Eve Balfour helped to institutionalize standards for organic agriculture by establishing the Soil Association in 1946. The association drew up its first set of standards in 1967 and published them as a four-page set of guidelines in the association's magazine, *Mother Earth*. A key principle of the standards centered on soil health: '[t]he use of, or abstinence from, any particular practice should be judged by its effect on the well-being of the micro-organic life of the soil, on which the health of the consumer ultimately depends' (Soil Association, 2013). The standards also prohibited routine use of anti-biotics in livestock and of industrial pesticides in organic agriculture. The standards for organic production used by the Soil Association were a codified set of principles used to evaluate the quality of an organic product by restricting synthetic soil inputs, and in many ways laid the foundational environmental principles that would make up components of the process- and product-based definitions of organic. This overt focus on soil's impact on the quality of organic food, and a lack of emphasis on other process-related activities, may have contributed to the uneven and diversely applied process-based principles that in later years would result in the co-existence of two competing definitions. Yet the Soil Associ-ation's standards were not entirely silent on all issues associated with the process-based definition of organic. It also attempted to protect legitimate organic producers and consumers from the damage to consumer confi-dence caused by fraudulent organic labels. In its early years, the Soil Association asked its members to register their farming establishments and sign a document stating that they would abide by the Soil Associ-ation's standards for organic agriculture (Vossenaar, 2003:12). Early efforts to regulate organic agriculture thus began as voluntary sets of standards maintained through self-monitoring. The principle of self-monitoring would become a foundational concept for many grassroots organic associations, which many early practitioners found hard to give up as various governments began to incorporate organic standards into regulation.

Building on the main principles of the Soil Association, more localized forms of organization emerged in some parts of the US. These were comprised of neighboring organic producers who knew, trusted and shared knowledge with each other. There was a degree of trust among organic producers and an assumption that association members were putting into practice many of the social and ecological principles associated with the process-based definition of organic. Trust through personal relationships was a very important component to the success of early organic agricultural associations. As Granovetter argues in his work on the importance of social relations in establishing trust in economic relationships, 'densely knit networks of actors generate clearly defined standards of behavior easily policed by the quick spread of information about instances of malfeasance' (Granovetter, 1985:419). In the early networks developing around the principles of the process-based definition, trust played an important role because many of the multifunctional qualities of organic products are what Hobbs (2001a:563) defines as 'credence attributes'. Unlike other 'experience attributes' of a product, such as color, shape or taste, credence attributes are things that are not detectable even after consumption. Since many of the attributes commonly associated with organic foods are process-based (for example, labor intensive, use of compost, etc.), and therefore undetectable in the physical characteristics of the food product, there is a need for quality assurance through some form of surveillance and enforcement of rules. Early grassroots organizations used mutual trust as a form of quality assurance. This was a reliable and effective form of quality assurance because organic practitioners, supporters of organic principles and consumers of organic products tended to have some type of personal relationship with one another. Early grassroots standards reflected the local conditions in which they were developed and accounted for factors such as soil type, available resources and topography, thus making them responsive to the needs of organic producers in a particular geographic setting (Berry, 1976:152).

There were also groups of organic producers and supporters who viewed organic agriculture as a way to challenge the social relations found in the mainstream culture of the 1960s, while remaining independent from government influence and corporate control. Some groups coming out of the counter-culture in California, Washington, British Columbia and Ontario decided to establish communes in rural parts of the country to live independently from what they viewed as oppressive social norms of North American mainstream culture. As Guthman explains, 'most of these communes practiced what were later codified as organic techniques, not necessarily by intention, but because self-sufficiency was a cornerstone of their ideology' (2004:6). Despite the claim that communes were able to escape social conventions and establish alternative forms of social organization, many of them were not free of the gender norms prevalent in conventional society at the time (Belasco, 1989). Because of the lack of consideration given to breaking down traditional gender norms in communes, many women questioned whether communes offered a legitimate alternative to the social relations found in conventional society.

Many women who were interested in organic agriculture and inspired by the second-wave feminist movement of the 1960s (which questioned traditional gendered divisions of labor and rejected the male-dominated communes) began to practice organic agriculture while applying socially progressive principles (Guthman, 2004). The growing interest and partici-pation of women in organic agriculture during the 1960s helped to infuse the goal of social sustainability into the developing process-based defin-ition of organic. In some early networks of organic practitioners increased attention was paid to addressing the socio-economic relations within agriculture, namely how work, profits and decision-making power were distributed on the farm. This drew attention to unequal gendered divisions of labor in agriculture but also issues of property ownership and the concentration of wealth within the agri-food sector.

Those early organic producers who decided not to establish alternative forms of collective living pursued privately regulated principles to determine acceptable practices in organic agriculture (for example, small-scale polycultural farms and non-chemical inputs). Most of the principles established through informal, grassroots associations were a reflection of what many organic producers were already practicing. The impetus for organizing associations around organic agriculture in Canada emerged as ideas from Europe made their way to North America. On a speaking tour in the early 1950s, German soil scientist Ehrenfried Pfeiffer came to Canada and discussed his experience with biodynamic agricultural techniques. Biodynamic farming techniques attributed to philosopher Rudolf Steiner emphasized the relationship between the soil, plants, animals and humans. It was the first method of food production in the midst of the dominant model of industrial food production to conceptualize food production as part of an interconnected system within nature. In 1924 at a lecture Steiner gave, he stated, 'nowadays people simply think that a certain amount of nitrogen is needed for plant growth, and they imagine it makes no difference how it's prepared or where it comes from... we've lost the knowledge of what it takes to continue to care for the natural world' (Steiner, 1924:9–10). Though not all organic practitioners used the biodynamic method, the holistic approach that Steiner and others championed resonated with those looking for alterna-tives to producing food that required fewer resources.

Like early organic agriculture, biodynamic farming had a strong ideological component. From the perspective of biodynamics, it was fundamental to see the biosphere as an interconnected, holistic system. Though aspects of biodynamics were criticized by some as unscientific, Pfeiffer's and Steiner's ideas about how to achieve balance and harmony among living things while producing food helped to inform the first generation of Canadians who decided to socially organize around alternative agricultural principles. The first formal organic agriculture organization in Canada was the Canadian Organic Soil Association (COSA) established in 1953 (Hill and MacRae, 1992:5). The COSA (later renamed The Land Fellowship) sponsored speakers so that they could travel across Canada to educate the public about organic agriculture. Christopher Chapman, a Canadian film-maker who headed the COSA, produced two documentaries on organic agriculture entitled *Understanding the Living Soil* and *A Sense of Humus* that were shown across the country.

Speakers who travelled across Canada also helped establish a number of organic farming organizations in the 1970s. In 1974, McGill University established its own Ecological Agriculture Projects program, which would become a clearing house of information on organic agriculture for Canadians. The following year, the Canadian Organic Growers (COG) was established by Peter McQueen who also established the Organic Gardeners and Farmers Association of Toronto. Originally, six Canadian provinces joined COG in the 1970s. In these provinces, conferences and meetings were held that served as spaces for practitioners to exchange information and gather knowledge about organic agriculture (MacRae, 1990). These were the first efforts to create a national network of organic practitioners and supporters across Canada. In 1977, COG began to publish its own newsletter called *COGnition* (COG, 2011).

The Establishment of Organic Certification Schemes

As organic food gained in popularity, more actors entered the organic sector to capitalize on its entry into the mainstream. Since many of the standards promoted by organic associations were based on self-regulation, the influx of increasingly diverse actors (who were often not part of the locally based associations) in the organic sector diminished the standing of trust as a form of quality assurance and led to collective action problems. The price organic products collect in the market is based on their credence attributes. Credence attributes, unlike experience attributes, can be subject to manipulation as goods that do not possess the claimed credence attributes enter the market and are substituted for the higher quality goods. This creates mistrust between producers and consumers which can shrink or dissolve markets. This is what Akerlof (1970) labels the 'lemons effect'. Since credence attributes are near impossible to verify based on visual inspection, it is difficult to determine whether a good is produced based on a set of agreed upon principled criteria without some type of enforcement mechanism. Non-compliance with the principles guiding a set of processes could happen by accident or intention. Either way, a trust-based form of quality assurance for goods valued for their credence attributes is problematic. As more actors enter the market, the trust-based system of quality assurance becomes ineffective and more difficult to maintain.

The early trust-based system of quality assurance for organic foods had the potential for profit-motivated individuals to free-ride off the self-regulatory nature of organic associations, while those committed to the substantive principles of organic agriculture voluntarily put those principles into practice (Ikerd, 1999). Many of the small, local organic associations had few enforcement mechanisms due to the nature of grassroots associations, relying on members regulating their own behavior and the strong role that mutual trust played in relationships among organic practitioners. Penalties for misusing the organic label were often unclear or non-existent, opening up the opportunity for those seeking to profit from higher demand for organic food to label their goods organic when they were in fact not produced based on the principles and standards of association charters.

Organic associations often had individualized standards that were not harmonized with neighbouring associations. But as markets for organic food expanded, the pressure to stretch out value chains across distances also increased and so did the need for reliable systems of quality assurance. This made it far more difficult to maintain personal trust-based relationships as a foundational component of assuring authenticity and justifying organic food's premium price. Because of the concern over the lemons effect, organic practitioners who wished to sell their products beyond local networks had to devise a quality assurance system to guarantee the credence attributes of organic food products.

Partially because of the potential for misuse of organic labeling and the ineffectiveness of self-monitoring, some informal networks began to transform into more organized associations that allowed for organic producers to institute more formal methods of monitoring. It was these private associations that transformed the informal principles that were part of a trust-based network into a formalized system of rules based on the enforcement of regulations regarding organic agriculture. The first organic agricultural association in the US was established in 1970 when a group of farmers from Vermont established the Northeast Organic Farmers Association (NOFA). The NOFA enabled organic producers and supporters of organic agriculture to come together under an umbrella organization. Knowledge was shared between members, and the NOFA eventually grew and split off into a number of local chapters (Henderson, 1998:17). The NOFA also began to grant organic accreditation and certification to its members on local and regional bases (Lohr, 1998:1126). Certification and labeling schemes were meant to assure buyers of organic products that a certain level of organic standards was met throughout the production process.

The certification of organic farms refers to the evaluation of an organic product to assure that it meets a specific set of criteria or standards. The establishment of certification bodies for organic products was by far the most significant element to the institutionalization of organic agriculture into private, sector-wide regulations. By making certification mandatory in a particular industry, only those who meet the certification require-ments can participate in the market (Stringer, 2006; Kaplinsky, 2000) thereby avoiding the market-eroding lemons effect. Certification is there-fore a successful way of addressing fraudulent organic claims and excludes those from the industry that do not meet the required criteria, because it addresses the collective action problem of past self-monitoring schemes. Certification in the organic sector today is quite rigourous. As Guthman explains:

> to be certified, growers had to fill out elaborate paperwork including a farm plan; agree to initial annual and perhaps spot inspections; fulfill whatever requirements for crop or soil sampling; pay various dues, fees and assess-ments; and of course, agree to abide by the practices and input restrictions designated by that agency and the law. (Guthman, 2004:129)

Many of the early requirements used for organic certification in the 1970s are still relied upon today.

The Rodale family established its own certification program in the early 1970s. This program awarded producers a seal of approval follow-ing independent laboratory tests of soil from farms claiming to be 'organic' (Belasco, 1989:161). In 1972, the Rodale family established an organic certification program that included 56 California organic grow-ers. The Rodale Institute also helped to establish the Oregon-Washington Tilth Organic Producers Association (Baker, 2004:1). CCOF emerged in 1973 from the program Rodale initially established. The CCOF began as a collective of 50 organic producers and consumers based in California (Dimitri and Richman, 2000:4). Its first newsletter published in 1974 includes a list of 13 requirements for organic certification (CCOF, 1974:16). These basic principles cover issues surrounding the use of the CCOF label, the required humus content in the soil and restrictions on using 'injurious' or 'harmful' materials in the production of certified organic food. Though these 13 principles were somewhat general and lacked an enforcement mechanism, they set the standard for certification schemes that followed. Some organic agricultural associations included substantive goals in their certification guidelines, in addition to regulations regarding allowable inputs and crop rotation. The California-based Farm Verified Organic (established in 1979), for example, includes standards for water conservation, labor practices and farm size in its certification schemes (Guthman, 2004:129). To become certified by Farm Verified Organic, producers must meet all of the standards set by the certifier. Farm Verified Organic now certifies producers, processors and handlers of organic products globally, including in the province of Quebec, and is accredited by both the Canadian and American organic certification programs.

Those in Canada who wanted to be certified in the 1970s often looked to third-party certifying associations. The certification bodies in Canada were established much later than their American counterparts and significantly later than similar establishments in Europe, given that the concentration of organic practitioners and supporters was much smaller. Although there were organic producers throughout Canada as early as the 1950s, there is little statistical evidence available showing how many people actually practiced organic farming in Canada at that time. According to the Canadian-based agriculture organization EarthCare, there were approximately ten organic farmers in Saskatchewan in the early 1970s (EarthCare, 2010). The first organic (and biodynamic) agricultural certifier to operate in Ontario was Demeter, which began certifying organic producers in 1982, 54 years after it was first established in Germany. Other provinces began to establish their own certifiers as the number of organic producers began to grow in the 1980s. The Organic Crop Improvement Association in New Brunswick was followed by Le Mouvement pour l'agriculture biologique (MAB) in Quebec, both established in the 1980s (Hill and MacRae, 1992).

Despite the emergence of small-scale standard-setting associations across states and provinces, the majority of states and provinces remained without state/provincial-level regulations (Amaditz, 1997). In the states and provinces without certification and standard-setting bodies, organic practitioners and consumers largely relied on the integrity of trust-based networks before the more formalized associations emerged. The establishment of private certifying associations met many of the needs of

organic producers and consumers as they established standards to evaluate the production processes of organic food. Since market competition
and output are central issues for organic producers in a globalized
market, maximizing efficiency and yields as well as guaranteeing quality
assurance are essential in order to compete with larger producers that can
meet economies of scale and reduce overhead costs through mechanization and more efficient labor practices (Jackson, 1998). Mutersbaugh, in
his study of organic certification systems, claims that 'monitoring systems ... introduce bureaucratic costs that rest heavily on producer
organizations and disrupt or differentially affect local governance and
economic management within producer organizations and villages'
(Mutersbaugh, 2002:1166). For this reason, certification schemes often
benefit actors who already have access to financial resources and whose
production processes are not linked to localized agri-food chains.

Over time, the emergence of a growing number of certifiers, some
state-sanctioned, others not, contributed to a complex web of rules,
regulations and standards that sometimes overlapped and were unevenly
enforced (Vossenaar, 2003:12–13). As early as 1971, the *Organic Directory*, a Rodale publication, cited the need for uniform standards: 'the
need to determine clear-cut, satisfactory standards as a measure of quality
represents one major first-order of business objective' (Rodale, 1971).
Others actors, however, saw no need for a separate certification standard
because 'organic foods are in no way detrimental to health ... therefore
there would seem to be no basis for any special regulations applicable
only to organic products' (White, 1972:32). Critics of the claims that
organic food is healthier and safer than conventional fare cited the lack of
a national-level certification system as a major problem within the sector.
An article published in a 1974 issue of *American Journal of Clinical
Nutrition* stated, 'there is no federal agency or law that defines and
supervises the label organic and certifies that such foods do in fact fit that
description; thus, there exists an avenue for obvious consumer fraud'
(McBean and Speckmann, 1974:1073). Although certifying associations
had the power to revoke certification if practitioners were found to
violate certification guidelines, these enforcements lacked legal structure
and formal enforcement mechanisms. Because there were no legally
enforceable regulations, it was difficult to determine whether producers
were actually meeting the standards set out by certifying associations.
Thus, the need for government intervention in organic agriculture
stemmed from the failure of self-regulation as cases of misuse of the
organic label multiplied, which threatened the integrity of the privately
sanctioned organic labels in the 1980s (McLeod, 1976:205; Rundgren,
2003:6; Vossenaar, 2003:14).

Organic Regulation Goes Public: State and Provincial Involvement

The growing number of organic producers across Canada and the US wanted to foster confidence in the organic label. This stimulated the development of organic certification schemes at state and provincial levels despite the continuing importance of local and regional organic associations. The move towards involving the government in enforcing regulations signaled a radical departure from the initial position of organic practitioners who insisted on autonomy from the state (Vossenaar, 2003:14). However, some in the organic agriculture sector argued that institutionalizing definitions for organic food and agriculture into government regulations could preserve integrity in relationships between producers and consumers, albeit not on a personal level. With public certification, organic products could be sold beyond the farm gate and the market for organic products could expand in a rigorous system of certification and regulatory enforcement. Having principles codified into law also created a closed market that allowed only those who met all the criteria to participate in the organic food marketplace.

Some supporters of organic agriculture maintained skepticism of government involvement, claiming that the orientation of governments towards agriculture was 'in direct opposition with the ideology of bioregional associations which subscribe to environmental principles of self-regulation, and local autonomy' (Berry, 1976:152). As Marsden et al. argue, rarely do policy makers have intimate knowledge of local practices and few are actually from farming backgrounds (Marsden et al., 1996:363). Thus, public regulation of organic agriculture began to divide those in the organic sector into one group that wanted government involvement to help expand markets and another that objected to the intrusion of governments into their private activities. There were fears that once government became involved in standards setting and inspections, the organic label would be under the control of individuals who did not have experience with organic techniques, or have intimate knowledge of the processes and practices associated with organic agriculture. Many organic practitioners wondered how organic standards would be incorporated into political institutions designed to support and promote conventional agriculture and its focus on economic efficiency.

The first foray into making private certification public was in 1973 when Oregon became the first state in the US to pass a law regulating organic food and agricultural processes in response to reports of fraud and inconsistencies in organic claims. In 1979, California passed its own Organic Food Act, which legally defined organic practices in California, although there were no provisions for support or enforcement. It was up to organizations like the CCOF to pursue instances of malpractice. Other states soon followed suit, but substantial differences in state organic farming regulation persisted across the US. Some states required third-party certification to determine whether a product met the standards of what constituted organic, while others did not. In some cases, *ad hoc* systems of self-monitoring by organic producers continued to be relied upon.

According to Amaditz (1997), the most serious problem facing the development of a coherent certification system prior to 1990 was that producers and marketers in 28 unregulated US states could make organic claims based on other states' definitions of organic practices, creating confusion for consumers. To solve this problem, individual states with growing numbers of organic producers and consumers began to establish state-level regulatory frameworks to police products labeled 'organic'. Individuals within a particular state can obtain organic certification through these state-level agencies that are monitored by state-level departments of agriculture. As of 2003, 14 American states with signifi-cant numbers of organic producers established state-level regulation and certification schemes (USDA/ERS, 2003). By 2012, there were 18 state-level USDA-accredited organic certifiers in the US under the National Organic Program (NOP), established in 2000.

Across the Canadian provinces throughout the 1970s and 1980s, there was little coordination between organic agriculture associations and governments. Certification programs developed on a much smaller scale and at a slower pace than in the US, mainly because there were fewer organic producers and a smaller market for organic products in Canada. Certification standards remained independent of provincial governments in Canada until the 1990s when Quebec instituted provincial regulations regarding the use of the term organic and organic agricultural techniques through the Conseil d'accréditation du Québec (CAQ). Organic producers in Quebec were then required by law to meet the standards set by the government of Quebec. In 1999, Quebec had a mandatory standard that was recognized by the USDA as equivalent to its own (Doherty, 2003). The CAQ has the authority to certify establishments producing organic products as complying with Quebec's provincial organic standards, and allows Quebec to have access to the US markets.

In 1993, British Columbia established the Organic Agricultural Prod-ucts Certification Regulation to establish a program to certify organic producers in BC (BC Reg. 200/93). The Certified Organic Association of British Columbia (COABC) administers organic certification regulation under the British Columbia Food Choice and Quality Act. The USDA recognized British Columbia's provincial regulations as equivalent to its own in 2003. British Columbia's interest in pursuing provincial standards that meet USDA guidelines reflected the fact that British Columbia had the highest proportion of land area dedicated to producing organic fruits and vegetables in Canada in the early 2000s (Parsons, 2004:3). Outside British Columbia and Quebec, provinces relied on private, third-party certification accredited by the Standards Council of Canada (SCC). Until the adoption of a national Canadian Organic Standard through the Organic Products Regime (OPR) in June 2009, third-party certification was the only option for Canadian organic establishments not located in British Columbia or Quebec. Despite the ratification of the national organic standard in Canada, producers in various provinces such as Quebec and British Columbia continue to be able to attain provincial level certification that indicates where the organic food product was produced above the requirements for national certification of organic products. For example, producers who want their products to carry the 'BC certified' organic label must adhere to the following principle of COABC's 'The Principles of Organic Farming': 'To allow everyone involved in organic production and processing a quality of life, which meets their basic needs and allows an adequate return and satisfaction from their work, including a safe working environment' (COABC, 2009:2). References to labor or the 'quality of life' of those involved in organic agriculture as well as other social aspects to the organic production process are not included in Canada's national organic standard, yet before the national standard was in place, provinces were free to include supplementary requirements for organic agriculture.

The impetus for expansion and formalization of certification for organic production processes was twofold: market actors wanted increased access to markets beyond specific localities, and consumers wanted quality assurance and access to a wider variety of organic products. Michelsen and Soregaard identify the paradox of organic agri-food policy entering the public realm in the EU context, which is applicable in the North American context as well: 'public and uniform certification systems seem to be paramount for the growth of organic farming. Hence, it seems to be a real paradox that organic farming must give up its self-rule and identity in order to obtain importance in agriculture' (2002:80). The public regulation of organic standards was difficult to avoid as producers and consumers of organic foods increased throughout the 1980s. The expansion of markets that extended beyond direct producer/consumer interactions was, in many ways, dependent upon regulations that could reliably be enforced to assure quality and integrity of the organic label. Although small, informal organic producers' associations continue to exist today, much of the authority over developing standards and regulations has been assumed by state and provincial governments. To truly expand beyond local markets, however, federal regulations were needed.

Institutionalizing Organic Food Regulation into National Policy Frameworks

As the global market for organic products in OECD countries continued to rise throughout the 1990s, the Canadian and American governments were under pressure to develop their own national regulations from those within the organic sector. As the market value for organic products grew, both federal governments took very different attitudes towards organic agriculture than they had in the past. As a representative from IFOAM noted at the 2001 OECD Workshop on Organic Agriculture, 'one of the main aims of establishing organic standards and regulations has been to foster trade in organic products' (Bowen, 2002). Despite this shared goal, in both Canada and the US, institutionalizing organic agriculture into federal agriculture policy frameworks proved difficult, and in some cases controversial, as government interpretations of the goals and standards of organic agriculture differed in some ways from those who practiced organic agriculture adhering to the process-based definition of organic. What has characterized policy processes in both countries is the debate as to whether national standards should reflect some of the principles of the process-based definition of organic.

Private actors representing both sides of the debate have played an important consultative role with both federal governments in devising national standards. The national policies for organic agriculture have been formulated to largely reflect the market approach to food production, and this is the primary reason why federal governments have involved themselves in regulating organic production processes. The most important role for federal level regulations is to facilitate the expansion of markets to export domestically produced organic products, though how organic has been instituted into national levels of policy making in Canada and the US differs in some considerable ways. The focus on market competitiveness in federal-level public policy led to convergence in terms of the majority of rules governing how Canada and the US regulate organic agricultural practices. The continued expansion of policy regimes for organic food and agriculture into national policy agendas has contributed to the marginalization of the original substantive and critical elements of the process-based definition of organic, in favor of the product-based definition favored by those subscribing to a vision of organic that is primarily based on the banning of chemical inputs, as well as biotechnology.

Regulating organic agriculture has been a priority for many governments in OECD countries like Germany, Sweden and Denmark since the 1980s, as their domestic markets developed at an earlier stage than those in North America (Tate, 1994:16; Michelsen, 2001b:4; Lockeretz, 2007). The self-regulation of organic practices that some organic farmers preferred because of its independence from state influence and control did not necessarily reflect the interests of all organic farmers, especially those who wanted to market their products abroad and those who felt their reputations were being tarnished by cases of fraudulence in inconsistently regulated organic food markets. Those who were critical of the lack of uniformity in private regulation in the organic sector encouraged the institutionalization of organic agriculture into government-enforceable regulations (Michelsen, 2001a:73; Padel, 2009). The tension between supporters of private regulation and supporters of public regulation is present in the case of the US and later in Canada as markets for organic products expanded in the 1980s and 1990s.

Some observers of the developing policy regime for organic agriculture claim that even though policies and standards for organic agriculture were integrated into existing national and global public policy frameworks, organic agriculture presents a challenge to the norms and principles governing public policy-making for conventional agriculture (Michelsen 2001b; Allen and Kovach, 2000). Others claim that policy standardization and the move toward global harmonization of public policy for organic food and agriculture is a symptom of the converging forces of economic globalization that has forced organic agriculture to conform to the norms and principles governing the global economy (Raynolds, 2000; Buck et al., 1997; Klintman and Bostrom, 2013).

American and Canadian organic policies exhibit some degree of convergence on the conventional model of organization applied to agricultural policies. However, the convergence of Canadian and American organic policies does not mean that both countries have instituted organic into policy frameworks in exactly the same way or included exactly the same things. As Bennett notes, 'often convergence is used as a synonym for similarity or uniformity ... [but] convergence [in activities] implies a pattern of development over a specified time period' (1989:219). Thus, while regulations regarding conventional and organic agriculture are far from identical, public policy governing organic agriculture has been fashioned with similar goals in mind: to incorporate it into global trade relationships to expand markets for organic products.

EXPANSION IN THE UNITED STATES

In the early 1970s, the USDA conducted a nation-wide survey of organic agriculture that revealed its rapid market growth and profit potential (Kuepper and Gegner, 2004:4). The USDA was looking for a competitive market solution to the farm crisis hitting America's 'corn belt' as a result of high energy costs brought on by the oil crises of the 1970s. Since organic agriculture used little or no petroleum-based synthetic inputs, it appeared to be a logical type of agri-food production for the USDA to investigate. In 1980 Washington University's Centre for Sustainable Agriculture and Natural Resources published *Recommendations on Organic Farming* that utilized the survey's findings. This report concluded that commercial organic agriculture could be competitive with conventional agriculture and could provide an alternative for conventional producers struggling to financially manage growing energy costs and farm debt. In an article entitled 'Organic Farming Becomes "Legitimate",' Carter (1980:254) comments on the contents of the report:

> The report ... does not suggest that a sweeping conversion of farmers to organic methods is either likely or desirable. But it suggests that many farmers can, and perhaps should, adopt organic farming practices, combining them with conventional practices if necessary or desired.

Referring to the same report as Carter, US Secretary of Agriculture Bob Bergland explained:

> [W]e think it is an important report – the first recent report to look at organic farming as a legitimate and promising technique. The past emphasis has been on using chemicals, but this has been driven by availability of low-cost oil. The economics of farming have now changed substantially. We now depend on imported oil ... and farmers are worried about these forces over which they have no control. (Carter, 1980:254)

Despite the promising tone of the report, the USDA could not take the position that organic agriculture was somehow a better option or produced 'healthier' or more nutritious food compared to conventional agriculture, considering that the vast majority of producers in the US practice conventional agriculture (Kuepper and Gegner, 2004). The USDA promoted organic agriculture and, at the same time, began to devise strategies to regulate organic agriculture, to 'assure consumers freedom of choice and to provide a niche market for strapped farmers. This allows for organic food to be grown and sold alongside conventional food without disparaging the rest of the food supply that the government must stand behind' (Guthman, 2004:164).

Although the market for organic products continue to be dwarfed by the market for conventional food products, the USDA's interest in the economic potential of organic agriculture grew, as did some organic practitioners' interest in devising a national set of standards to govern organic production processes. Because of the growth in size of the domestic organic market in the US in the 1980s, the development of a US national organic standard served the interests of domestic consumers who wanted a recognizable, uniform labeling system of quality assurance, and it served the interests of some organic producers eager to expand their operations and markets for their products across the US and abroad. By having a formal set of regulations pertaining to organic agriculture, organic producers could become certified under a national label that would assure consumers of the authenticity of purchased organic products, and would be recognized by the international community. A USDA label on all organic products that met a formal, nationalized set of regulations would allow more domestic American organic producers and processors to export their products abroad and to sell their products as 'certified organic' in other states.

In 1984, the Organic Foods Production Association of North America (OFPANA) was formed to lobby the US federal government to create regulations regarding the organic sector. The OFPANA consisted of producers, shippers, retailers, distributors, exporters and importers of organic products. The OFPANA was a small and private lobby group that would later become the OTA, and continued to play an important consultative role in shaping federal regulations in the US and Canada in the 2000s. From its inception, the OTA has argued that harmonized, national standardization eliminates the overlap of organic production standards at the sub-national level. Harmonized national standards provide consumers with a reliable label to evaluate organic products that are imported from abroad (OTA, 2006). The OTA is also highly supportive of efforts to increase the ease at which organic goods are incorporated into longer value chains. There was also growing public pressure to establish a US-label for certified organic products. The *New York Times* (*NYT*) reported in 1987 that, 'unfortunately, not all food labeled "organically grown" actually is, and it is almost impossible for consumers to distinguish between the genuine and the fraudulent by looking. The industry is working on a verification system that will certify producers who adhere to certain principles of organic farming' (Burros, 1987). Another journalist following the progression of the organic movement notes, 'one serious problem in the marketing of organic foods is the lack of uniform standards and certification processes' (Marter, 1989). In an interview with *NYT*'s reporter Keith Schneider, a manager of a large organic farm in California stated, 'the Federal Government is going to be forced to make rules on organics uniform, and the sooner the better ... every state has a different set of rules. The longer it goes on the more uncertainty there will be among growers and consumers' (Schneider, 1989).

Just after the Organic Foods Production Act had been introduced into Congress by Senator Patrick Leahy of Vermont, *Organic Gardening* magazine contributor Steve Daniels heralded the proposed legislation as a positive step toward formal recognition for organic foods: 'if passed, this bill will revolutionize the way America farms ... The organic act is landmark, revolutionary legislation ... The federal government has finally recognized the importance of organic' (Daniels, 1990:7). Observers of the changes in organic food regulation in the late 1980s and early 1990s tended to focus on the positive environmental benefits organic regulation would have for the reduction of pesticide use in agriculture and residues on food. There was little mention in popular periodicals at the time of the linkage between organic agriculture and some of the more traditional principles of social responsibility and scales of production associated with the organic movement of the 1960s.

Lobbying by environmentalists, organic producers, consumers and organic associations, such as the OTA, played a significant role in the inclusion of the Organic Food Production Act (OFPA) in the 1990 US Farm Bill (USDA, 1990; Tick, 2004). The OFPA was the first effort by the US federal government to institutionalize some of the principles of organic agriculture embodied in state-level certification schemes, such as allowable inputs. Although some in the organic agriculture community like the OTA applauded the recognition organic agriculture received from the US federal government, others were skeptical of how organic policy would be incorporated at the federal level and what would happen to state-level certification schemes. Would they have to conform to a less stringent national standard, or would state-level certifiers be able to have differentiation among certification requirements?

Numerous advocates of the process-based definition of organic questioned the product-based form of quality assessment and expressed concern regarding the impact of national policies on the capacity of organic agriculture to maintain a critical alternative to conventional modes of production, such as large-scale cultivation. Supporters of the process-based definition feared that federal level regulations would only emphasize the technical, material elements of organic foods and leave out many of the substantive elements of organic agriculture such as scale of production. Some US organic producers feared that federal regulations would water down more localized, practitioner-devised standards developed over decades in order to be applicable to a wide variety of organic products being produced in a wide variety of circumstances. As one critic of the impending US regulations for organic agriculture said, 'the pragmatic approach to organic growing is pushing our ethics into obscurity and irrelevance' (Urwin, 1986:10–11).

As the drafted US national standards continued to develop in the late 1990s, concerns were raised with the standardization that federal regulations would create, and that bureaucratic red tape and high certification costs would prevent small-scale organic producers from meeting those regulations and attaining certification. Gene Logsdon, a Wisconsin organic dairy farmer, expressed doubts about the move to standardize and institutionalize organic agriculture into federal level policies: 'the concern that many of us have is that in the process of putting organics into a regulatory framework, we might produce a conceptual model which is so complex that we face prohibitive costs and intimidate farmers with unmeaningful rules and paperwork' (Logsdon, 1993). Ronnie Cummins, the executive director of the OCA, argued that the USDA deliberately 'watered down standards' to privilege powerful industry, while ignoring the needs of small-scale organic farmers and growers in the US (Lilliston and Cummins, 1998).

As in other economic sectors where business and government work together to develop policy, the 'privileged position' of business interests was evident early on in organic policy-making processes (Lindblom, 1977). In 1992, the National Organic Standards Board (NOSB) was established to create a national organic program for the US and to supply the USDA with suggestions for what the National Organic Standard (NOS) should include. The NOSB has a diverse membership and includes producers, handlers, processors, retailers, environmentalists, and consumer groups (Coleman and Reed, 2007). In 1997, the NOS was presented by the USDA with notable exemptions in the proposed legislation from some recommendations of the NOSB. Particularly, as supporters of the process-based definition had feared, the USDA opted to not include many substantive goals associated with organic agriculture (Nestle, 2004:232).

The drafted NOS legislation regarding organic standards did not restrict the use of GMOs, bio-sludge or irradiation from the technical definition of what products could carry the 'organic' label. The USDA was sensitive to the position of the conventional agri-food sector that regulations for organic agriculture should not include language that negatively portrays other forms of agri-food production, as well as the claim that irradiation and GMOs did not alter the material aspects of an organic product, and therefore should not be included as banned inputs and processes (Nestle, 2003:233). But due to the massive public outcry and protest (over 275 000 complaints), the USDA was pressured to include in the NOS regulations that products under the USDA's certified organic label could not contain GMOs, bio-sludge or be subject to irradiation (Baker, 2004:2). Despite the attempts by some conventional agricultural producers to influence the content of NOS, concerned practitioners and consumers of organic products pressurized the USDA to regulate all inputs and processes that change the material aspects of organic. But critics in the conventional agri-food sector claimed that the USDA decision to ban the use of GMOs and bio-sludge in the NOS and organic production processes was politically motivated, and based on something other than scientific evidence (Nestle, 2004:233). The amended OFPA, which included the NOSB's recommendations to ban certain substances and processes, was implemented in 2002 and covers all organic 'cultivated crops, wild crop, livestock, livestock feed, and handling (preparation and processing) operations' (Riddle and Coody, 2003:52).

The NOP was formally established in December 2000. Producers who wished to be certified and sell their products as USDA-certified organic had to comply with the NOP by 21 October 2002. The NOP is based on one piece of legislation. The OFPA (7 U.S.C. 6501–6522) covers handling requirements, certification, labeling, accreditation of certifying agents and allowable substances related to organic agriculture. The OFPA created a standardized set of criteria to judge all organic products grown in the US to provide the consumer with a guarantee of authenticity of the products purchased, and established an accreditation agency under the NOP. Accreditation refers to the authority given to certifying associations to certify producers, processors and handlers as 'certified organic' establishments allowed to label their products as 'USDA-certified organic'. The USDA accredits certification agencies that certify organic firms. Firms wishing to use a USDA-certified organic label on their products must be certified by an accredited agency. Examples of such agencies are CCOF, Oregon Tilth, and the Organic Crop Improvement Association. The USDA also accredits state-level certifiers like Pennsylvania Certified Organic as well as foreign agencies to grant certification such as Canadian-based Pro-Cert. This NOP's system of accrediting certifiers builds on the infrastructure of certification created by third-party certifiers in the 1970s as well as state-level programs.

The NOP and the OFPA regulate the use of the label 'organic' and certify products that fall into one of three categories: '100 percent organic', 'organic' or 'made with organic ingredients' (USDA, 2002d). The USDA frames the OFPA as a 'marketing label', not a 'code of conduct', and is focused on consumer protection from false 'organic' claims (Bostrom and Klintman, 2006:164). From the position of the USDA, the NOP is not meant to include any substantive principles often associated with the process-based concept of organic. Individual states are able to have their own organic programs that may apply higher supplementary voluntary standards to the operations they certify (for example, bird-friendly, fair trade), but at the very least they must comply with the standards and regulations established by the NOP.

The role of the NOP as an overseeing/accreditation agency has influenced the product-based nature of the regulations and standards for certified organic in the US. The primary drive from stakeholders involved in the deliberative process was to create a nationwide certification system that recognized a type of food production that did not use synthetic chemical inputs as a minimum requirement for certification. This created a policy baseline, or a 'policy floor', for the 'certified organic' label. This system of regulation and management had to take into consideration that many types of farming (livestock, crops), and many forms of production (monoculture/polyculture, large-scale/small-scale), would be covered by the NOP and therefore could not include specific regulations that may be covered by other agencies (such as the Department of Labor or pre-existing USDA regulations or the EPA). As long as production met the minimum requirements set out by the NOP, certification of an operation would be granted, but this does not stop operators from doing more than what the 'policy floor' establishes. They can do this by attaining voluntary certifications, such as eco-labels like 'Carbon Neutral Certifi-cation' or distinctions related to the treatment of animals such as 'Animal Welfare Approved'. This model builds in more choice on behalf of organic practitioners to conduct their business as seen fit anywhere in the country so long as they abide by the minimum standard set of rules for inputs established by the NOP. It also creates differentiation in the market (and some would argue confusion regarding food labels), which increases choices for consumers who wish to purchase products with multiple certifications (mandatory and voluntary) including socio-economic or ecological standards. The model of certification that is in use via the NOP serves farm practitioners of all sizes across the country, providing universal standards with flexibility for 'add-on' certifications if desired.

Despite the inclusion of the NOSB's recommendations to ban the use of GMOs and other chemical inputs, the OFPA makes no mention of labor standards or issues of ecological sustainability that are associated with the process-based definition of organic. Not all of the supporters of a national standard were pleased with the final draft of the OPFA. In an interview, OCA's Ronnie Cummins criticizes the final OPFA because it 'say[s] nothing about subsidized water, animal treatment, labour standards or food miles' (Ruiz-Marrero, 2004). The exclusion of some substantive environmental standards (for example, water usage, farm size) and the entire exclusion of social standards (for example, fair treatment of labor) were exempt from the final OFPA (Guthman, 2004:117). For example, under the Applicability Preamble of the NOP in the 'Changes Requested But Not Made' section is point 10: 'Fair Labor Practices on Organic Farms'. Although some contributors to the creation of the NOP were eager to include labor standards in the actual NOP legislation, labor practices are not considered integral to regulations pertaining to organic agriculture, because 'other statutes cover labor and worker safety standards' (USDA, 2002c). As Guthman points out, however, there is little or no reference to labor in any enforceable regulations pertaining to organic agriculture, and labor as a distinct interest group has been virtually left out of the policy process (Guthman, 2004:182). The primary focus, as expressed by observers of the changes occurring in the early 1990s was on validating a form of food production that did not use chemical inputs or biotechnology, which some perceive as a healthier and more nutritious alternative to conventional foods (Burros, 1987).

Another issue demonstrating the lack of inclusion of substantive principles found in the process-based definition of organic in the NOP is the absence of regulation pertaining to farm size. As previously discussed, farm size is a pertinent issue to organic producers subscribing to the process-based definition because small, polycultural farms have a greater ability to foster biological diversity and recycle farm wastes. In the NOP, there are no restrictions on certifying large-scale organic farming operations. As stated in the NOP, 'the final rule does not contain such a prohibition [on factory farms] because commenters did not provide a clear enforceable definition of "factory farm" for use in the final rule' (USDA, 2002d:93). Under the NOP, regulations regarding the size of organic farms are not included because the size of the farming operation does not affect the qualities of the end, material organic product. The lack of distinction between scales of farming establishments demonstrates the absence of formal recognition of the importance of what happens throughout organic agricultural processes in the USDA's organic regulations.

Organic producers who oppose the lack of definition of farm size included in the NOP, such as a small-scale organic producer from US-based Marquita Farms, argue that 'the federal standards are just about what "thou shalt not do". It doesn't talk about what you should do: soil conservation, reducing the distance food had to travel, staying away from monoculture' (Mark, 2004). The regulations are meant to 'level the playing field among producers' while securing market access for economically competitive producers. As DeLind notes, the institutional-ization of the technical aspects to organic agriculture and the 'lack of specific definition allow[s] many ... to associate [organic food] with important characteristics of scale, locality, control, knowledge, nutrition, social justice, [and] participation' even though actual production pro-cesses may not display these qualities (DeLind, 2000:200). But unlike the public outcry that followed the original draft of the NOS and OFPA for excluding inputs and processes that altered the material components of an organic product, the exclusion of substantive social and ecological goals associated with the process-based definition of organic received far less public acknowledgement by the organic sector. The desire to establish a nationwide set of standards and guidelines to make sure the rapidly expanding domestic growth in the organic sector as well as imports coming into the US were monitored for authenticity, safety and quality trumped efforts to factor in substantive social and ecological principles. Further, diversity among organic producers and advocates in terms of their own interests and beliefs in regulating the social aspects of organic production contributed to a lack of consensus on broader issues. How-ever, efforts have been made to bring organic practitioners and supporters together to develop a vision for organic now that it is part of the regulatory system. The National Organic Action Plan 2010 developed over five years by the Rural Advancement Foundation International outlines a policy agenda and strategic goals for the future of organic agriculture in the US. Its purpose is to maintain 'organic integrity' in the midst of structural changes in the sector and rising consumer demand. It calls for an organic policy agenda that 'reflects the broader social, environmental, and health values of the organic movement and the associated benefits of organic food systems afford society. The goal of the NOAP project is to *establish organic as the foundation for food and agricultural production systems across the United States*' (original emphasis, Hoodes et al., 2010:5). The policy objectives include ensuring a fair marketplace for small, medium-sized, and family farmers and workers, and enhancing access to organic foods for people of all income levels (Hoodes et al., 2010:9). Despite the clear preference for the product-based definition of organic in the US regulations, stakeholders supporting the process-based definition continue to work towards influ-encing policy so that it is more reflective of the substantive values and goals traditionally associated with organic food.

US government financial support for organic agriculture increased steadily after the passing of the NOP. In 2002, the US Farm Bill included over $32 million (USD) devoted to organic agriculture. $13.5 million (USD) was earmarked to financially support producers seeking organic certification, while $15 million (USD) was for research on organic agriculture. The remaining $3.75 million (USD) was devoted to promoting and developing markets for value-added US organic products (OTA, 2002). The 2008 Farm Bill increased funding for organic agriculture to approximately $112 million (USD) over five years. The 2008 Farm Bill boosted funding for all three of the areas of organic agriculture as in the 2002 Farm Bill, but also included $5 million (USD) to collect and distribute data on organic agriculture in the US (Willer and Kicher 2009:226). This has made statistics on organic agriculture in the US some of the richest and easily accessible in the world through the USDA's National Agricultural Statistical Service (NASS) and the ERS. In 2014's Farm Bill, the USDA committed $11.5 million (USD) to organic certification cost-share assistance, easing the financial burden of certification for small-scale producers and handlers seeking certification (USDA, 2014).

In 2009, the NOP became an independent program within the USDA, receiving its own Deputy Administrator (before 2009, the NOP was under the control of the Agricultural Marketing Service (AMS)). But because of the increasing expansion of organic markets, and the corresponding need for regulatory oversight, it was necessary to establish the NOP as part of the USDA. Since 2009, the NOP continues to expand in terms of the funding it receives and infrastructure. It now accredits over 100 organic certification agencies, 44 of which are based outside of the US, and two of which are based in Canada (Canadian Seed Institute and Pro-Cert). This represents the formal recognition of organic agriculture and its fledgling institutions as a legitimate form of food production. It also represents the formal linkage of conventional and organic agriculture under the same federal department.

EXPANSION IN CANADA

The ratification of the NOS in the US played an important role in how Canada shaped its own national organic standard. Since 1999, Canada had a set of voluntary standards that were not supported by formal regulation. Canada's national organic standards attend to some issues of process associated with the process-based definition of organic such as ensuring the humane treatment of animals (CAN/CGSB-32.310-2006: iii). Yet, much as in the case of conventional agriculture, a significant portion of the Canadian standards were very similar to the principles found in the US's NOP. The US is Canada's leading trading partner in organic products, but Canada does not have the same level of influence over the US as the US has over Canada. Since Canada exports a significant proportion of its agricultural products, it has to respond to the demands of importers to some degree. Canada's lack of a ratified national standard until 2009 was an ongoing trade issue for EU importers of Canadian organic products, as it was argued by EU regulators that a Canada-wide standard would make it easier for importers to understand which parts of the standards were equivalent to their own national standards. Statistics Canada reports that the primary motivation behind the establishment of a Canadian national organic standard was 'to meet these nations' [EU] standards' (Wunsch, 2003:187). Although Canada had an informal national standard since 1999, the Canada OPR was not ratified into law until June 30, 2009 (COG, 2014).

Several Canadian governmental agencies began exploratory research into organic agriculture in 1999 when the EU fully implemented its organic standards [EC no. 2092/91] (EC, 2012). In 1991, the Canadian Organic Unity Project (COUP) was formed to develop a regulatory system to govern the production and handling of organic agricultural products in Canada, as the export potential for organic products was recognized (Doherty, 2003). The Canadian Organic Advisory Board (COAB) replaced the COUP in 1993 and held a consultative role with the Canadian General Standards Board (CGSB) in developing national standards in 1997 (CGSB, 1999). The CGSB drew up a lengthy outline of the proposed Canadian National Standards for Organic Agriculture (NSOA) regulations, while proposed outlines of standards were debated between the CGSB and the COAB. The CGSB consulted with committees from the organic sector such as the Standards Committee on Organic Agriculture, which represents producers and interest groups like the OTA. Deliberations continued throughout the early 2000s (AAFC, 2014; OTA, 2005).

Though a voluntary standard existed in Canada since 1999, the CGSB routinely revisited the standard as the market for organic products in Canada evolved. In 2002, AAFC requested that the CGSB revise the standards to facilitate trade with major trading partners like the US, Japan and the EU. Existing standards already in place in these listed countries along with the Codex's 'Guidelines for the Production, Processing, Labelling, and Marketing of Organic Food' were considered in the revisions. The 2006 version of the standards includes both trade consider-ations and adherence to established guidelines from Codex (PWGSC, 2013). The standards were amended three more times (2008, 2009, and 2011) to reflect changes to technical requirements and additions of substances on the permitted substance list. The Canadian Food Inspection Agency (CFIA) also required changes be made to the standards in these three rounds of revisions. Coordination between multiple agencies in Canada (CGSB, AAFC, CFIA, Foreign Affairs and International Trade Canada, etc.) has contributed to the length of time it has taken to establish and implement the standards.

Canada's decision to institutionalize standards and regulations for organic food and agriculture at the federal level moved much more slowly compared to the US, but like the US, it included a number of interest groups in the policy process. The *ad hoc* Organic Regulatory Committee (ORC) was made up of private sector actors and suggested in 2003 that a Canadian organic standard should consist of a federal regulation, a national organic standard and maintenance system, an optional national symbol, competent authority, network of organic certifi-cation bodies, surveillance and enforcement system, advisory body, national registry and funding arrangements (ORC, 2003:7). The ORC included representatives from national and regional organizations as well as certifiers and businesses with an interest in implementing a national organic standard. Representatives from all areas of the organic sector were consulted in the drafting of a Canadian National Standard for Organic Agriculture (CNSOA) including the Canadian Organic Growers, Organic Crop Producers and Processors Ontario, Organic Trade Associ-ation, Organic Livestock working group and a member of the Animal Welfare Task Force (ORC, 2003). The OTA, as well as the COG and Organic Federation of Canada (OFC), worked with the CGSB to provide some of the resources to convert the voluntary standards to mandatory processes as well as administrative support during the deliberative process.

In discussions between the CGSB and private sector representatives in the ORC regarding the Canadian organic standard, the consensus between provinces in developing this regulatory standard was paramount. If provinces did not wish to be responsible for regulating organic practices themselves, the ORC suggested that they might delegate jurisdiction to the federal government (ORC, 2003:6). Canada's ratified national organic standard recognized both Quebec and BC's provincial standards as equivalents (CFIA, 2003). Like the standards set by the NOP in the US, the COR act as a baseline set of standards for organic production within the respective countries. As in the US, practitioners in

Canada who seek certification from agencies with principles relating to social responsibility and restricted off-farm inputs can voluntarily apply higher standards. Also, as in the US, there are diverse ways in which organic agriculture is practiced on farms throughout the provinces in Canada. Where the product-based definition of organic is most visible is in the national organic standards of both countries that primarily focus on monitoring and regulating what goes (or what does not go) into the organic food product. Though there are elements in both sets of regulations that attend to elements of process, such as animal welfare in Canada, evidence of a strong lean towards the product-based definition is clear in both cases.

Canada has largely fashioned its organic food and agriculture policies considering the scope of the US model. The motivation behind Canada's effort to harmonize its policies with the US increased the flow of exports to the US market by making cross-border transactions easier, as demonstrated in the revisions focusing on policy harmonization that led to the 2006 version of the standards. The Canadian and American organic standards were deliberately similar in scope and content as the model of standards and certification used by the USDA was developed in tandem with Canadian standards. But, much like in the development of the US's NOP, there were divisions among the organic producers as to what should be included in the national standards and what should not. The absence of a national organic standard for Canada presented a serious barrier to trade for producers in the Canadian organic sector, and its implementation was necessary to assure consumer confidence in Canadian organic products. Those involved in the deliberations over Canada's organic standard recognized the important role that harmonizing standards played in passing mandatory legislation but also saw the value in an enforceable, national standard. In an interview with *alive* magazine in 2005, the Canadian Organic Initiative Coordinator of the Certified Organic Associations of BC said:

> initially, there was a need for export access, mostly to the European Union ... Subsequently, the more important need is for consumer assurance and protection ... we want mandatory regulation that would retain the aspects of the systems operating already. We want the government to produce a legal overlay, to provide surveillance. (quoted in Hancock, 2006)

The CNSOA, as put forth by the CGSB and like the USDA's NOP, includes no comments or regulations regarding conditions for workers on organic establishments, or mention of the social principles associated with the process-based definition of organic (USDA, 2002d), though some elements of process are included in the standard. The final draft of the Canadian Organic Production System (COPS) refers to organic agriculture as 'based on principles that support healthy practices [that] aim to increase the quality and the durability of the environment through specific management and production methods. They also focus on ensuring the humane treatment of animals' (CGSB, 2006:7; CAN/CGSB 32.310-2006). This is a significant part of the national standard. It requires herbivore livestock to have access to pasture during the grazing season (weather permitting) and access to 'open air' (CAN/CGSB 32.310-2006: 6.1.3). Battery cages are prohibited, though there are not specific limits on the size of farms. The involvement of such groups as COG, the organic livestock working group and Animal Welfare Task Force played an important role in getting animal welfare components into the final version of the standard. The CNSOA only pertains to criteria necessary to certify crop and livestock production, handlers, transportation and labeling. Although the COPS standards address some of the technical environmental principles of organic agriculture like banning synthetic pesticides, it does not address farm size or consideration of labor conditions that are commonly associated with organic agriculture practices.

Supporters and practitioners of the process-based definition of organic have been critical of the policy process regarding a national standard for Canada, both from inside the deliberation process and as observers. COG, for example, took issue with the phrase 'organic product' that appeared in section 1 of the revised standards before 2006. In a letter commenting on the proposed legislation in 2006, COG representative Laura Telford stated, '[t]he phrase "organic product" inappropriately places the emphasis of the regulation on the final organic product. The Canadian organic standard, CAN/CGSB-32.310-2006, is a process standard that governs the processes used to produce and process organic products.' COG recommended that the phrase 'organic product' be replaced with 'organically produced product'. In the 'Introduction' to the standard the revised version states, 'This standard is intended for certification and regulation to prevent deceptive practices in the marketplace. The certification of a process rather than a final product, demands responsible action by all involved parties' (CAN/CGSB-32.310-2006, Introduction). Though the process is what is being certified, the product is what is evaluated by importers and consumers based on whether it carries an organic label reflecting national certification. As previously stated, many consumers buy certified organic foods for the qualities they embody, not necessarily for the processes used to achieve said qualities.

Some Canadian organic producers did not support a harmonized national standard, claiming that it would render existing programs redundant and de-legitimize small-scale farmers' certifications at the regional and provincial levels who could not afford another set of certification (Welling, 1999:61). Some organic producers have opted out of national certification standards partially as an act of political protest to what they view as state sanctioned regulations devised to control the activities of organic producers and to impose heavy certification costs upon them (Seiff, 2005).

In June 2009, Canada finally ratified its own national organic standard. The OPR is part of what is referred to as the Canadian Organic Regime (COR). The CFIA is the enforcer of the OPR through the Canada Organic Office, which was created with the ratification of the OPR. The COR includes a set of mandatory national standards, consistent labeling rules, a national logo and an enforcement and oversight platform through the CFIA. There are three pieces of legislation that make up the COR: Organic Production Systems – General Principles and Management Standards [CAN/CGSB-32.310]; Organic Production Systems – Permitted Substances List [CAN/CGSB-32.311]; and Organic Products Regime [SOR-2009-176] (CFIA, 2012).

The Canadian Organic Standards and Permitted Substances List were made mandatory for all organic food and livestock feed sold or imported into Canada with the ratification of the OPR. It is maintained by the CGSB through its Technical Committee on Organic Agriculture. At the sub-state level, provincial organic standards and certification programs remain intact, and several provinces such as British Columbia and Prince Edward Island continue to invest in and develop programs to support their own organic sectors, but province-level programs and funding remain uneven across the country. In regards to labeling, as part of the OPR, strategies were purposely harmonized with the US and the EU, though there are some notable differences. Canada, for instance, does not allow for a '100% organic' claim to be made on any labels while this is permitted under US regulations. Canada also does allow for the 'Transition to Organic' claim to be made on certified organic labels, unlike the EU (Willer and Kilcher, 2009:244). One of the most intriguing parts of the ratified national organic standards for Canada was the inclusion of regulations that allow the CFIA to enter into equivalency arrangements with its trading partners, namely the US, EU and Japan (OPR, SOR/ 2009-176: Part 4, Sec. 27 (1b)). This clearly indicates the desire to move towards a degree of regulatory convergence among trading partners and the desire to build regulatory flexibility into the Canadian organic market activities.

Proponents of the Canadian national organic standard made the claim that not having a national organic standard disadvantaged Canadian producers domestically and globally, and added to the cost Canadian consumers were paying for organic food products because of multiple certification requirements (Klonsky, 2000:234). But it is unclear how federal level regulations benefit domestic production for domestic consumption in Canada, when the guiding principle included in the OPR is to facilitate the movement of Canadian-grown organic agricultural products across its borders. Before the OPR was ratified, Canada imported most of the processed organic goods sold in Canada from the US, including soybeans, fruit juices, frozen vegetables and dried fruit. Sales of processed organic products make up almost 90 percent of all organic products sold in Canada (Gold, 2005; USDA, 2005:3). In terms of domestic consumption, according to Macey's 2004 report to AAFC, 62 percent of organic produce, 60 to 85 percent of grocery, and 10 percent of dairy products purchased in Canada are imported from the US (Macey, 2004:26).

The AAFC reported that exports of Canadian organic products were worth over $63 million (CAD) to the Canadian economy in 2003 (Kortbech-Olesen, 2004:5). Canadian exports reached approximately $390 million (CAD) by 2011 (Willer and Lernoud, 2013:71). Most of Canada's current organic production is for export rather than domestic processing or consumption (USDA, 2005:2). The US imports 42 percent of Canada's organic exports, which mainly consist of organic goods minimally processed such as produce, grains and seeds, while the majority of the rest are destined for the EU and Japan (Macey, 2007:8). Concerned organic producers groups have blamed the large volume of organic imports from the US to Canada for the slow growth of organic farming in Canada.

Canadian Organic Grower's executive director Laura Telford has expressed concern over the fact that Canada imports the vast majority of its organic products. She argues that the flood of imports gives Canadian farmers little incentive to convert to organic farming because of the associated start-up costs (Stephenson, 2007:9). Though having an enforceable national standard was essential for the Canadian organic sector to expand, its ratification is unlikely to change the reliance Canada has on US imports, since the national regulations were devised to facilitate trade, not promote domestic production for domestic consumption.

The vast majority of certified organic agricultural goods grown in Canada is not processed, sold or consumed in Canada. This has not changed in any significant way since the ratification of the OPR. Though the OTA and the COG have launched campaigns to promote and encourage Canadians to consume more Canadian-grown certified organic foods, the primary focus of policy appears to be focused on the promotion and expansion of international markets for Canadian organic agri-foods. Much of the economic growth in the organic agricultural sectors in Canada is a result of growing export markets rather than domestic demand. It is not yet clear if the OPR will boost domestic consumption of certified organic products in Canada, or if policy initiatives in Canada will continue to focus on nurturing the expansion of international markets for Canadian certified organic products.

The year 2009 was historic for the regulation of organic agriculture in Canada and the US, ushering in a new era for bilateral trade in organic agricultural products when on June 17, USDA Deputy Administrator Barbara Robinson and CFIA Director of Agri-Food Division Jasper Komal signed the Canada–US Organic Equivalency Arrangement. The organizing principle of this equivalency arrangement is that signatories may have different regulations, and differing regulatory frameworks and means of enforcement and inspection, but so long as a similar policy objective is achieved, organic goods are treated as equivalent and not required to undergo repeated inspections and/or certification (CFIA, 2012).

Canada's list of conditions regarding certified organic foods from the US is much longer than that of the US's conditions for Canadian certified organic foods. For example, US agricultural products produced with the use of sodium nitrate (also used as a food preservative commonly found in cured meat products like bacon) cannot be marketed as organic in Canada (as stated in the CGSB's Organic Production Systems – Prohibited Substances List [CAN/CGSB-32.311] and the Canada – US Organic Equivalence Arrangement), whereas Canadian-produced animal products that are derived from animals treated with antibiotics cannot be marketed as organic in the US (CFIA, 2012). This equivalency arrangement highlights some of the differences between the regulation of organic food in Canada and the US, but the guiding principle embedded in this document is clearly a mutually shared one. The agreement institutionalizes the product-based mode of assessment in bilateral trade of certified organic products, despite the fact that some process elements are included in national standards. Though at the domestic level, national organic standards differ in Canada and the US, their international policies share the goal of facilitating market expansion for domestically produced organic food products.

CONCLUSION

The primary motivator for institutionalizing organic agriculture into national level public policy in the late 1990s and early 2000s was to expand markets for organic products beyond local and domestic levels. Certifying agencies functioning with provinces and states continue to set higher standards for the operations they certify, and in some cases include process-based elements of the process-based definition of organic. National standards have not prevented operators from practicing organic agriculture according to higher standards, but they have created policy baselines that primarily focus on farming inputs and how they alter the end, material product. Higher standards required by third-party certifiers are beneficial to consumers who want to purchase organic products that are produced with adherence to social and ecological goals not included in the national standards. But different certification standards at the sub-state level (that may or may not include process-based assessment criteria) creates an uneven playing field and tiers of certification that further complicate the definition of organic and may prove confusing for consumers trying to support progressive types of organic production. Certified organic foods in some instances carry a confusing array of labels that the average consumer would find difficult to define and differentiate without an intimate knowledge of the nuances of the regulations, standards and guidelines for labeling and an awareness of the intricacies of how governmental agencies interact and make decisions regarding agriculture and food. The effort to assure quality and enforce standards has helped to streamline production standards, but to the average consumer, various (and abundant) labels on organic foods have added another layer of confusion.

5. Globalizing organics: the role of trade agreements and international organizations in regulating trade in organic food

INTRODUCTION

Although the production standards and labeling guidelines covering organic food have only been formalized since the late 1990s, regulations covering organic food have been part of the international policy discourse on food and agriculture for over 20 years. Organic food made its global policy debut when in 1992 Finland notified the WTO Committee on Technical Barriers to Trade (TBT of its 'Draft Decree on Indications Referring to Organic Agricultural Production for Foodstuffs'. The notification states that products can only be labeled 'organic' if they are subject to 'inspections and surveillance under the Decree' (Finnish Ministry of Trade and Industry, 1992). The purpose of the notification to the Committee on TBT was to establish a set of standards for organic products produced in the European Economic Community (EEC), or imported into the EEC, as there was little reference to organic food in the trade policy prior to Finland's notification. The decree became Council Regulation (EEC) no. 2092/91 and continues to be the set of standards and regulations for organic agriculture in Europe. The European guidelines presented to the Committee on TBT signaled the official entry of organic food into the global trade regime. However, as far back as the 1940s, with the establishment of the Soil Association in the UK and the emergence of the International Federation of Organic Agriculture Movements (IFOAM) in the 1980s, organic agriculture received considerable attention from the international community as an alternative to industrialized food production. As the profit potential of globally traded organic products became realized, new actors such as governments and corporations joined the international community in promoting organic agriculture. As organic food was incorporated into globalized markets it became subject to the same trade agreements as any other traded agricultural or food product, which has significant implications for putting the process-based definition into practice.

This chapter focuses on how the principle of 'like' products embedded in various trade agreements and in trade-related institutions and organizations plays a crucial role in how the qualities of globally traded organic agri-food products are defined. The first section examines the North American Free Trade Agreement (NAFTA) and the World Trade Organization (WTO), and how the liberal principles of trade they embody apply to organic products circulating through the global trade regime. The second section looks at two trade agreements administered by the WTO: the Agreement on TBT and the Sanitary and Phyto-Sanitary Measures Agreement (SPS). It shows how the limitations these agreements place on the inclusion of process in defining the qualities of traded agri-food products privilege the product-based definition. This section also addresses the issue of Processes and Production Methods (PPMs) in global trade disputes, citing some cases pertaining to agri-food. It discusses how PPMs apply to globally traded organic agri-food products. The third section takes a closer look at the role that three international organizations have in setting standards for organic agriculture and regulating the movement of organic agri-food through the trading system: the Codex Alimentarius Commission (Codex), the International Organization for Standardization (ISO)[1] and the IFOAM. Through the analysis of how trade agreements institutionalize the product-based approach to producing, processing and distributing organic agri-food by pushing for harmonization and equivalency of policy, this chapter shows how the central principles of the process-based definition of organic are largely undermined as organic agri-food becomes part of the global trade regime.

BASIC PRINCIPLES OF TRADE: NAFTA AND THE WTO

Scholars studying the inclusion of organic food in the global trade regime in the early 2000s such as Dabbert (2003) claimed it would only increase the profile of the organic sector and strengthen its association with food quality and safety in the global marketplace. Lohr and Krissoff (2000) and DeLind (2000) similarly held the view that the recognition and

---

[1]   'Because International Organization for Standardization would have different abbreviations in different languages, it was decided when the organization was established to use a word derived from the Greek word "isos" meaning "equal"' (www.iso.org/iso/en/aboutiso/introduction/index.html). 'ISO' is the established acronym and is used here.

harmonization of policies at the global level would encourage the expansion of the markets for organic food worldwide. Trade agreements such as NAFTA and the General Agreement on Tariffs and Trade (GATT), administered by the WTO, function to ease the movement of goods and services across borders by removing and limiting discriminatory barriers to trade. As organic products enter the global trade regime, they are subject to the same rules of trade as all other food and agricultural products covered by those agreements. This section shows how the principles and scope of NAFTA and select WTO-administered agreements apply to organic agri-food. Although neither NAFTA nor WTO-administered agreements specifically address organic agriculture, the legally enforceable rules that privilege liberal principles of trade have significant implications for organic food as it moves through the global trade regime.

NAFTA

NAFTA was signed in 1994 by Canada, Mexico and the US and is built upon the Canada–United States Free Trade Agreement (CUSFTA) signed in 1989 between Canada and the US. In the 1980s, the US preferred to enter into bilateral trade agreements out of frustration with the speed of multilateral trade negotiations (Schaeffer, 1995:255). Both agreements share the objectives of eliminating barriers to the trade in goods and services, and removing restrictions on foreign direct investment between its member states. NAFTA emulates many of the core principles found in the proceeding GATT agreements administered through the WTO.

The primary purpose of NAFTA is to lower trade barriers to trade and investment between the three signatories. Article 102, Chapter 1 of NAFTA states that one of the primary objectives of the trade agreement is to 'eliminate barriers to trade in, and facilitate the cross-border movement of, goods and services between the territories of the Parties' (Government of Canada et al., NAFTA, 1994: Chap. 1, Art. 102). In essence, NAFTA is premised on creating a regionalized market with few restrictions on the movement of goods, services and investment. By signing onto NAFTA, member states have encouraged the opening up of domestic economic sectors to privatization, deregulation and subjected certain economic sectors to the harmonization of policy (Kratochwil and Ruggie, 1986:757). In NAFTA's preamble, the text states that the governments of Canada, the US and Mexico resolve to reduce distortions to trade and enhance the competitiveness of firms in global markets (NAFTA, 1994: preamble). Regulations that enforce social or environmental standards in trade-oriented sectors that create differentiation among member states are considered to be discriminatory barriers to the free flow of goods, services and investment (Cohen, 2007). However, there are two 'side agreements' that accompany NAFTA: North American Agreement on Labor Cooperation (NAALC) and the North American Agreement on Environmental Cooperation (NAAEC). These side accords were put in place to address the concern over the 'race to the bottom' towards minimal environmental and labor standards that some countries might implement in order to attract investment within North America. The accords commit each government to enforce their own labor and environmental laws, and do not commit them to any new ones. NAALC and NAAEC allow for a government or member of civil society to file complaints and have them investigated if governments are not enforcing their own labor or environmental laws (Allen, 2012; Commission for Environmental Cooperation (CEC), 2000). Though some incremental policy change can be attributed to the process that is part of the CEC, the concerns raised by the complainants are only marginally addressed through the process. Indeed, there are channels to address the 'race to the bottom' fears raised by civil society groups that accompanied the signing of NAFTA, but the institutions put into place have had limited effectiveness since the creation of the NAAEC.

How a good or service can be produced in the most efficient and effective way is the favored condition as instructed by the legally binding text of NAFTA. In this regard NAFTA is based on the principle of National Treatment as found in the GATT. It reads, 'with respect to a state or province, treatment no less favorable than the most favorable treatment accorded by such state or province to any like, directly competitive or substitutable goods, as the case may be, of the Party of which it forms a part' (NAFTA, 1994: Chap. 3, Art. 301). The principle of National Treatment states that competitive and substitutable 'like' products must be treated the same whether they are domestically produced or produced by foreign firms outside of a member state's borders. A like product is defined as 'a product which is identical, i.e. alike in all respects to the product under consideration, or in the absence of such a product, another product which, although not alike in all respects, has characteristics closely resembling those of the product under consideration' (The Anti-Dumping Agreement, Art. 2.6; Subsidies and Countervailing Measures Agreement, Art. 15.1, fn. 46). Where the product is produced, who produces it and how it is produced cannot be used as legal grounds to deny access of a product to a member country's domestic market if it is determined to be competitive and a suitable substitute for a similar product.

Chapter 11 of NAFTA also extends the principle of National Treatment to investment. Art. 1102 of Chapter 11 states that "[e]ach Party shall accord to investors of another Party treatment no less favorable than that it accords, in like circumstances, to its own investors with respect to the establishment, acquisition, expansion, management, conduct, operation, and sale or other disposition of investments' (NAFTA, 1994: Chap. 11, Art. 1102). By extending the principle of National Treatment to investment, the inclusion of a number of social or environmental regulations tied to foreign investment is limited because they would restrict many foreign corporations from investing in various sectors of a domestic economy. NAFTA applies to the trade of most agri-food (with a few exceptions such as eggs, poultry and dairy) between the US, Canada and Mexico, and to sugar and syrup products between the US and Mexico (NAFTA, 1994: Chap. 7, Annex 703.2, Sec. A and B). Section B of Chapter 7 of NAFTA is based on the principles of the SPS Agreement, which is administered by the WTO (NAFTA, Chap. 7, Art. 712, 'Basic Rights and Obligations').

Specific to trade in agricultural goods, regional agreements like NAFTA have been shown to influence domestic policy options available to member states and to encourage the convergence of policy outcomes. In Josling's (2001:190) study of regional trade agreements and agriculture, he shows that the regional integration of agriculture production networks is part of a larger effort to harmonize regulation at the multilateral level. The global integration of agricultural policies is facilitated by regional institutions, which include the liberalization of agricultural sectors in their agreements. Schaeffer (1995:259) claims that by deregulating activities of agribusiness, regional institutions further contribute to the transnationalization of agricultural value chains and promote the 'monopoly power of TNCs'. Despite not being explicitly mentioned in its text, organic products that cross borders in North America are subject to the rules and regulation of NAFTA. The equivalency agreement covering organic food between Canada and the US (discussed in Chapter 4) reinforces the trade principles embodied in NAFTA. Product equivalence is the focus of these agreements as well as NAFTA, not equivalence of production processes. Though NAFTA does not deny the ability for a process-based interpretation of organic to exist, or deny consumers access to purchasing organic products that are produced based on standards higher than the baseline requirements for national certification, it does reinforce a set of production standards that only meets the baseline requirements for certification and product equivalence. It also puts downward pressure on any type of agri-food trade policy to conform and adhere to the principles of liberal trade that it is based upon. Organic firms participating in global markets must be competitive, and in many cases this means subscribing to the principles of the product-based definition of organic, as producers in competing countries may not be subject to the same requirements (beyond restricting synthetic inputs and GMOs) as domestic producers in Canada or the US.

NAFTA encourages market expansion to capitalize on international demand, and most of the public policy at the national level in Canada and the US follows the logic of producing for export markets. Canada orients most of its organic production towards satisfying the US market, while US production is directed towards supplying domestic demands but also exporting to Canada, the EU and Japan (USDA, 2002a; Macey, 2004). The transnationalization of organic food value chains is evident from the volume of organic products exported from Mexico to the US and from the US to Canada. In 2008, 74 percent of all organic food products sold in Canada were imported from the US while Mexico exports almost all of it organic produce to the US and Europe (ACNielson, 2009; Sligh and Christman, 2003:17).

By participating in the regional trading bloc which privileges production that employs the product-based definition of organic that does not include the social costs to production, practitioners of the process-based approach to organic agriculture find themselves at a market disadvantage if their products enter foreign markets. As Raynolds notes in regard to harmonized certification requirements, 'organic certification appears to reassert industrial and commercial quality conventions, based on efficiency, standardization, bureaucratization, and price competitiveness' (Raynolds, 2004:10). Because price is a major factor in a competitive market with varieties of organic products to choose from, cheaper organic products traded among NAFTA member states or circulated through regionalized value chains have a market advantage over organic products that voluntarily internalize social and environmental costs passed onto the consumer. Large-scale organic operations can reduce the price a consumer pays by achieving economies of scale and not adhering to process-based principles that may add to the final cost, such as fair labor practices. In a 2008 survey conducted by Whole Foods Market, North America's biggest retailer of organic and health foods, two-thirds of consumers preferred to buy organic (and natural) products if the price was comparable to non-organic products (Willer and Kilcher, 2009:232). Since there are no limits on who can invest or participate in organic value chains, large-scale actors that choose to subscribe to the product-based definition of organic are privileged in NAFTA and can provide cheaper organic products to a growing market by either regionalizing value chains or purchasing successful, smaller firms and integrating them into pre-existing transnationalized value chains. Process-based practitioners are still able to put substantive principles into practice, but often struggle to remain competitive against operations in the mainstream market that can offer cheaper priced certified organic products that adhere to baseline certification requirements.

The WTO

Considering that the Uruguay Round negotiations used the legal texts of NAFTA as a model, many of the benefits extended to transnational business in NAFTA are also a part of multilateral agreements adminis-tered through the WTO, including the Agreement on Agriculture, which the WTO enforces. Until the ratification of the Agreement on Agriculture, agriculture and trade in food products were not part of multilateral trade agreements. Though agricultural trade has some degree of regulation imposed upon it at the international level, it continues to be treated differently from other products. Under the GATT system previous to the Uruguay Round (1986–94), there were many difficulties in applying the rules and principles of trade evenly as there was no formalized adminis-tering body or mechanisms for appealing rulings. Often more powerful members of the GATT would ignore agreements that they did not benefit from, while weaker members had little choice but to participate for fear of being excluded from the global trade regime (Wolfe, 1998). The Uruguay Round's main objectives were to address the uneven application of agreements among members, to address newer trade issues such as services, investment and intellectual property and to develop a compre-hensive agreement on trade in agricultural products. The Uruguay Round would be referred to as the 'single undertaking' – meaning signatories had to agree to all of the agreements presented throughout the round with several exceptions for developing countries (Wolfe, 1998:94).

The WTO was created to develop, administer and enforce the GATT as well as other agreements pertaining to services, investment, intellectual property and agriculture among member states. It was also created to develop more comprehensive agreements through successive trade rounds. As Hoekman and Kostecki (2001:1) state:

> [T]he underlying philosophy of the WTO is that open markets, non-discrimination and global competition in international trade are conducive to the national welfare of all countries. A rationale for the organization is that political constraints prevent governments from adopting more efficient trade

policies, and that through the reciprocal exchange of liberalization commitments these political constraints can be overcome.

The WTO offers members mechanisms for resolving trade disputes when they arise either through the dispute settlement mechanism, the application of safeguards or the implementation of a member's negotiated exceptions from WTO agreements. Although the hope in creating the WTO's Dispute Settlement Body (DSB) was that it would alleviate the number of trade disputes, this has not been accomplished (McMichael, 2004). The Appellate Body as part of the DSB was created to hear appeals from members that dispute settlement rulings on trade disputes. This decision-making body can overturn, modify or uphold rulings of the Dispute Panel. Once the appellate report is issued, its contents are binding and must be adopted by all parties in the dispute. Issues related to food and agricultural goods continue to remain a major area of trade disputes between countries. Agricultural issues were a major contributing factor to the collapse of the Doha Round of trade negotiations in 2007.

Despite the success of the WTO in institutionalizing liberal principles of trade into enforceable international law, the rules continue to be applied unequally and unevenly. The more economically and politically powerful a member state is (and therefore able to demand exemptions from aspects of trade agreements), the more likely it is to use the principles and rules of the WTO to its advantage. In some cases, more economically powerful member states do not adhere to dispute settlement rulings and choose to pay fines for violating rules of trade or apply non-tariff barriers to imports rather than modify trade policy. For example, Brazil brought a case against the US government's subsidization of US Upland Cotton producers to the WTO DSB in 2002. The DSB ruled in favor of Brazil (and the Appellate Body upheld the ruling in 2005), yet the US refused to abide by the conditions of this ruling. Still unable to reach an agreement after five years, in 2010 Brazil and the US agreed to formulate a framework to find a mutually agreed solution to the dispute. As long as this framework is in effect, Brazil has agreed not to impose countermeasures approved by the DSB (WTO, 2013b: Dispute DS267).

Despite its shortcomings in successfully solving trade disputes involving food and agricultural goods, the WTO wields significant power in demanding policy harmonization between participating members and standardization of the criteria used to judge the integrity of tradable goods and services. Policy harmonization is also required for national organic standards and regulations. Some observers have heralded the move towards policy harmonization for globally traded organic agri-foods promoted by the WTO, claiming it 'reduce[s] information asymmetries along the marketing channel from producer to consumer' while reducing the costs passed onto the consumers through harmonized certification schemes (Lohr, 1998:1125; Lohr and Krissoff, 2000:212). The diverse standards of localized organic production networks have been replaced with harmonized, global standards as organic food production methods and products are subject to the rules of the global trade regime (Mutersbaugh, 2005). To date, however, the move toward mutual recognition of equivalence has been more successful than outright harmonization.

In many of the agreements formalized through the Uruguay Round, the end product is recognized as the only aspect of the production process that can be used to restrict the movement of goods across borders, which reduces the ability for a member state to discriminate against a product from another member state based on aspects that do not affect the qualities or safety of the product. This principle is a foundational component of equivalence agreements for organic products signed by trading partners. Harmonizing national trade policies and equivalency agreements, as promoted by NAFTA and the WTO, restricts the role which 'process' plays in distinguishing the characteristics of 'like' products and works against the logic of the process-based definition of organic. Organic foods circulating in the global trade regime are subject to the same rules and principles as other food products.

THE TBT AGREEMENT, THE SPS AGREEMENT AND PPMS

As with NAFTA, internationally traded organic agri-foods fall under similar legal jurisdictions of the WTO as other agricultural products. The revised TBT Agreement administered through the GATT and the SPS Agreement (part of the Agreement on Agriculture negotiated through the Uruguay Round) are relevant to organic agri-foods in the global economy. Both the TBT and the SPS 'concern the application of technical measures, food safety and animal and plant health regulations' (OECD, 2003a:119). The TBT and SPS are based on technical *regulations* applied to the material aspects of the end product whereas *standards*, which usually contain issues of process, are considered substantive aspects of evaluation. Both agreements allow member states to develop standards for goods that go above and beyond technical regulations but 'in the case of standards, non-complying imported products will be allowed on the market, but then their market share may be affected if consumers prefer products that meet local standards such as quality' (OECD, 2003a:120). This means that products meeting the lowest common regulation must be given the same market access whether or not they are sourced inside a country's economy, although local products may have a market advantage because of consumers' belief that their national standards are higher than those covering imported products of similar material quality (Pedersen, 2003:246). SPS and TBT share the principle of achieving harmonization of standards and equivalency in food control systems. These standards are based on scientific evidence of appropriate measures to achieve and monitor the safety of globally traded food products.

The TBT Agreement

Many of the regulations regarding food safety and animal and plant health were part of TBT, created during the Tokyo Round of GATT negotiations prior to the creation of SPS (Stanton, 2004). As a result of the Tokyo Round, ending in 1979, 32 GATT members signed on to a multilateral agreement called the Standards Code or the Agreement on Technical Barriers to Trade. Up until the end of the Tokyo Round, technical barriers to trade were the most widely used Non-Tariff Barriers (NTBs) exporters encountered. An NTB is defined as 'any governmental device or practice other than a tariff which directly impedes the entry of imports into a country and which discriminates against imports, but doesn't apply equal force on domestic production or distribution' (OECD, 2003a:41). The conclusion of the Uruguay Round of trade negotiations in 1994 strengthened and clarified the scope of TBT (WTO, 2014b).

TBT covers 'all products including industrial and agricultural prod-ucts' (WTO, 2012a; TBT, Art. 1.3). It also covers regulations, standards, testing and certification procedures that facilitate the free movement of goods across borders but it does not allow for what the WTO considers 'unwarranted protection for domestic producers' (OECD, 2003a:8; WTO, 2012a: TBT, Art. 5.1.2). The primary objective of TBT is to prevent the use of technical requirements to create 'unjustified' barriers to trade. Specific to agri-food products, TBT covers standards related to quality assurance of food products except for those requirements covered by SPS. Specifically related to agri-food, TBT covers quality provisions, labeling, nutritional requirements, packaging and product content regu-lations. It requires member states to use existing international standards to encourage policy harmonization, but there are special provisions made for developing countries and security interests (WTO, 2012a: TBT, 'Preamble'). TBT emphasizes that a member state should attempt to harmonize its standards to internationally established ones but it does not specifically mention the standards-setting bodies that should be referred to as those which set benchmarks. Most importantly, TBT strongly promotes the idea that if two sets of standards exist that achieve the same objective, the least trade-restrictive set of standards should be followed. As of 2003, only one trade dispute was raised regarding organic labeling under TBT. The details of the trade dispute have not been made public, but the majority of trade disputes regarding labeling have stemmed from the claim that labels have been used by a member state to create a barrier to imports that discriminate against a like product from another member state based on a higher set of standards (OECD, 2003a:9).

The SPS Agreement

Unlike TBT, SPS specifically covers trade issues regarding health and food. SPS gives member countries the right to apply measures to protect the health of humans, animals and plants. It deals specifically with 'rules for food safety, and animal and plant health standards' (WTO, 2012b: SPS Measures, 'Introduction', 1998). The text of SPS defines sanitary measures as rules that can be 'specific process criteria, certifications, inspection procedures, or permitted use[s] of only certain additives in foods' (OECD, 2003a:132). The primary purpose of SPS is to make sure that member states are not applying health and safety measures to imported agri-foods to restrict access to domestic markets in an inconsistent, arbitrary or unjustifiable manner. Member states are expected to treat imported products similarly to products produced domestically where similar conditions prevail (WTO, 2012b: SPS Art. 2, para 3). Member states are to base their national sanitary and phyto-sanitary standards on international measures established by the Codex, discussed in greater detail in the next section. Stricter measures may be applied by a member state if scientific evidence demonstrates that an imported good poses a risk to human, plant or animal health.

SPS regulations are based on risk assessment grounded in scientific evidence; that is, regulations regarding food safety must be based on science that the SPS Committee has widely established. All WTO members must accept other WTO members' standards as equivalent to their own, as long as they are based on scientific evidence. But what is to be considered 'scientific evidence' has proven more difficult to ascertain than the regulation suggests. The principles embodied in SPS have garnered criticism from those who claim they deny member states the ability to discriminate against importing products they determine to be dangerous to human, animal or plant health that other member states do not. Varying interpretations of what constitutes scientific evidence was the primary concern for France when it banned the import of hormone-treated beef from the US and Canada (WTO, 2014a: DS26). The scientific evidence presented by France indicated that hormone-treated beef was hazardous to human health, while the US and Canada claimed there was no scientific evidence that proved it was unsafe for human consumption (Hoekman and Kostecki, 2001:196). The US claimed in 1996 that the EU's refusal to import hormone-treated American and Canadian beef was in violation of several GATT articles (Article III, or XI, SPS Art. 2, 3 and 5), TBT Article 2 and the Agreement on Agriculture's Article 4 (WTO, 2014a: DS26). The panel reviewing the trade dispute ruled that the EU was in violation of Articles 3.1, 5.1 and 5.5 of SPS. The EU requested that an appellate body review the case, but the appellate body upheld most of the panel's findings (the EU was in violation of Articles 3.1 and 5.5). Despite this ruling, the EU stated that it could not comply with the appellate body's findings, and the trade dispute over EU imports of hormone-treated American and Canadian beef continued until 2009. In 2009, the US and the EU signed a memorandum of understanding concerning the import of beef not treated with particular growth hormones. Duties were increased by the US to certain products imported from the EU. Since 2013, the EU and the US have revised the memorandum as of October 2013 (WTO, 2014a).

The case of hormone-treated beef sheds some light on the difficulties involved in defining what constitutes scientific evidence, and how it can be interpreted differently among member countries when it comes to food safety and human health. Rulings that undermine a state's decision to deny the entry of goods that are deemed to be harmful or pose a risk to human health threaten the ability for states to enact food safety regulations, for fear they will be challenged at the WTO. The issue of providing 'scientific evidence' to support trade-restrictive food safety standards related to organic agriculture may present a future challenge to including environmental principles related to soil health in national organic standards and certification schemes.

PPMs

A principle of trade that both TBT and SPS share is opposition to trade restrictions based on 'non-product related' PPMs that do not shape the material characteristics of a final product (WTO, 2012e: 'Labelling').

Using PPMs as grounds to deny the access of a foreign-produced good into a domestic market is in direct conflict with the principle of National Treatment as outlined in GATT, Article 3 (WTO, 2013a; WTO, 2014b). A member state cannot use activities occurring during the production process in a foreign country to deny entry of its products into the domestic market unless it can scientifically be proven to damage human, animal or plant health. However, some PPMs are acceptable grounds to discriminate against a product under SPS such as the use of child, prison or forced labor or the trade of endangered species (according to the Convention on International Trade in Endangered Species of Wild Fauna and Flora (CITES) agreement). Process standards are quite different from product standards. Product standards refer to the outcomes of a domestic party using a product, while 'process standards are meant to control negative environmental by-products of the production process in foreign countries' (US Congress, 2005:149). Production processes occurring in another country are difficult for an importing country to monitor and cannot be legally used to deny the access of an import into a domestic market under WTO rules (US Congress, 2005:149). The inability for member states to consider PPMs when importing products has a number of implications for environmental standards as they relate to food.

One of the most notable trade disputes related to PPMs emerged in the 1990s with the 'dolphin-tuna case' between the US and Mexico and other complainants (WTO, 2011). The US wanted to deny the access of Mexican-caught tuna because the way it was harvested entangled and killed dolphins, which violated the US Marine Mammal Protection Act. If it cannot be proven that the way in which the fish were caught did not violate the act, the act states that all fish imports from the country are to be embargoed by the US. The embargo also applied to 'intermediary' countries that processed the tuna before it reached the US, such as Canada, Italy and Japan. Mexico lodged a complaint in 1991 and the GATT Dispute Settlement Panel ruled that the US was indeed in violation of the GATT. The panel stated that the US could not embargo tuna from Mexico based on 'the way tuna was produced' but it could apply regulations based on the 'quality or content' of the imported tuna. It also decided that including a 'dolphin-safe' label did not violate the GATT because it was meant to prevent deceptive advertising practices. The panel report was circulated but it was not adopted, and the US and Mexico settled their trade dispute 'out of court' (WTO, 2011). The 'dolphin-tuna case' drew attention toward trade issues pertaining to production processes, and to the 'product versus process' issue.

In 2012, the DSB of the WTO decided to adopt the panel and appellate body's decision to overturn the ruling in the 'dolphin-tuna case'. The appellate body found that the panel did not use the correct approach in upholding the decision that agreed with the US's 'dolphin-safe' labeling requirements, determining that they were not unfair technical regulations. The appellate body found that regulatory distinctions based on fishing methods were deemed to unfairly restrict market access to tuna imported from Mexico that did not meet proper certification standards because the certification was only required for tuna caught in the Eastern Tropical Pacific Region (where Mexico sources its tuna) and not for tuna caught outside of this region. This technical regulation under the 'dolphin-safe' certification was a legally enforceable condition that had to be met in order for exporters to gain access to the certification 'dolphin-safe' label and therefore to be able to export to the US. The requirement for certification that grants access to the US domestic market was deemed by the appellate body to be in violation of the TBT as the standard used to deny market access based on the method of capture, but not equally required of tuna entering the US market caught in other regions.

The recent and dramatic change in the dolphin-tuna case and the ruling of the 'product versus process issue' speaks to the blurring lines between voluntary standards, mandatory technical regulations and PPMs in the trade of food products. The overturning of the DSB ruling has set a precedent for including process methods in technical regulations. As Trujillo (2012) states, 'it suggests that any legislative or regulatory act that affects market access and contains legally mandated and enforceable conditions may constitute a technical regulation' because the appellate body considered a 'non-product-related' PPM as a discriminatory technical regulation. As found in both TBT and SPS agreements, the inclusion of PPMs in standards are to be restricted because of the concern that they could be disguised trade barriers, but this issue remains contentious among member states. As a report issued by the US Congress notes, 'a central issue with respect to PPMs is whether [WTO] laws can differentiate between different goods based on the processes or methods used in their production, if those processes or methods are not reflected in the observable and measurable physical characteristics of the product itself' (US Congress, 2005:149).

The issue of PPMs in global trade is thus a major obstacle for those who would like to see environmentally damaging production processes (for example, clear-cutting) used in foreign countries as grounds to deny access of their products to an importer's domestic market in the hope that those processes will eventually be phased out. Supporters of the WTO's position on denying the inclusion of PPMs as grounds for restricting entry of a foreign good to a domestic market claim that eco-labeling (based on what occurs during the production process) is in violation of

WTO agreements, particularly TBT (Hobbs, 2001b:272; Jacobsen, 2002:11). The WTO continues to struggle with environmental concerns presented by member states and how environmental standards should, if at all, factor into trade policies of member states; also how process should be interpreted in regulation covering agri-foods (Kerr, 2001:63).

Implications for Trade in Organic Products

Though there has only been one instance of issues surrounding the trade in organic products being brought to the WTO DSB, it is clear that restrictions based on the inclusion of process in trade agreements influence the development of national regulatory standards and how trading partners choose to deal with the global expansion of organic markets. As markets continue to expand for organic food, market actors from around the world will continue to be interested in participating. But those who wish to put into practice the environmental and labor considerations associated with the process-based definition of organic may be put at a market disadvantage as the global markets for organic agri-food continue to expand.

Regarding environmental considerations, the WTO created the Trade and Environment Committee in 1994 to flag environmental issues in need of further discussion among trading partners. This committee discusses trade issues around environmental requirements set by member states and works on ways of avoiding the use of these types of requirements as protectionist barriers to imports that may not meet the requirements of the importing member state, particularly focusing on how these issues affect developing countries. One of these issues is the growing presence of organic food and products circulating in the global trade regime. At a committee meeting in 2007 developing countries including Kenya, India and China voiced that the growth of the global organic market could be beneficial to their domestic economies, but took issue with the uneven national standards, environmental restrictions and private voluntary standards in developed organic markets such as in the US and the EU, as well as with the lack of a harmonized international standard (WTO, 2012d).

Despite the lack of an international harmonized standard, major trading partners exporting and importing organic products have recognized the value in equivalence recognition, for example the Canada–US Organic Equivalency Arrangement signed in 2009 (see Chapter 4), the Arrangement between EU and Canada on Equivalency in Organic Products in June 2011 and the US–European Union Equivalence Arrangement that took effect in June 2012. The Canada–EU arrangement is the result of lengthy negotiations between the trading partners. Both Canada and the EU scrutinized each other's production and control systems and determined them to be equivalent. Through this arrangement, both nationally certified organic labels are recognized as legitimate in both trading partners' domestic markets. The US–EU arrangement establishes mutual recognition of the trading partners' organic production rules and control systems as equivalent. Organic products certified in the US or the EU are legally allowed to be sold and labeled as certified organic in both countries, which eliminates the need for duplicate certification to US standards or vice versa (European Commission, 2012a, 2012b, 2012c).

These bilateral arrangements have the distinct purpose of recognizing the equivalence of certification requirements to eliminate or reduce duplication of inspections. On one hand, this is a positive step towards reducing regulatory overlap and therefore transaction costs which can add to the final price passed on to the consumer. On the other, the ratification of equivalence arrangements among Canada, the US and the EU, and the move towards harmonization in requirements, certification and standards, may act as further barriers to developing countries eager to gain access to these domestic markets. Bilateral harmonization agreements for trade in organic products could be a source of future trade disputes between WTO member states as markets continue to grow for premium-priced organic goods. The inclusion of environmental standards in national-level regulatory frameworks for organic agriculture, though not as substantive as those found in the process-based definition of organic, may be viewed as protectionist, and technical barriers to trade by developing countries eager to access growing foreign markets.

Although the treatment of labor and issues of farm size are fundamental to keeping economic relations embedded in the social relations that are associated with the process-based definition of organic, under SPS, issues such as how labor is treated in the production processes and the size of organic farms are not allowed to be included in national regulations. Since there is no 'scientific' basis for including fair labor standards and farm sizes as enforceable parts of national regulations for organic food production, including more substantive goals in national regulatory frameworks is almost impossible for WTO members. Thus, unsafe practices for workers (such as farm workers hand weeding or using a short-handled hoe) in food production cannot be used as reasons to deny access because the working conditions do not compromise the safety of the food – only the safety of the worker. The short-handled hoe was banned in 1975 in the US because it was determined to be the cause of many back injuries among farm workers in California. Hand weeding is used instead, which has also been proven to cause serious injuries among farm workers. Currently, legislation in California bans hand weeding, but organic farmers are exempt from the regulation because of organic agriculture's dependence on manual labor. Getz, Brown and Shreck (2008) explore the complexities of California legislation banning some forms of manual labor on farms, including organic farms. In this case study, some small-scale organic farmers objected to the banning of these practices because the effort was perceived of as representing the interests of agribusiness (less manual weeding, more dependence on industrial farming technologies).

If a production process does not directly affect the qualities of the end product, it cannot be used as grounds to deny entry of a foreign good into a member's domestic market. Although member states have the capacity to pursue domestic policies that promote 'technical regulations and conformity assessment procedures', labor and environmental standards are not included in what the WTO technically considers 'organic'. The WTO itself does not define organic, but relies on the definitions provided by other international standards-setting bodies like Codex and IFOAM (OECD, 2003a:8; Daugbjerg, 2012:56). Members have little recourse against imported organic products that may not meet their social and environmental standards put into practice by domestic producers. This can disadvantage domestic producers who voluntarily put organic principles into practice by adding to the final cost of the product, which may make their products less competitive than imported ones.

INTERNATIONAL AUTHORITIES ON ORGANIC AGRICULTURE STANDARDS: CODEX, ISO AND IFOAM

TBT and the SPS state that harmonization of international standards is possible if all members attempt to base their regulations upon pre-existing, well-established ones, such as those established by the ISO.[2] This section examines the role that standards-setting organizations such as the Codex, ISO and IFOAM have in creating organic standards and the integration of organic food into the global trade regime.

---

[2] See footnote 1 on the definition of ISO.

Codex

WTO member states developing organic standards are strongly advised to do so in coordination with other standards-setting bodies like Codex and ISO, both of which are recognized by TBT as authorities on international standards setting (WTO, 2012a: TBT, Annex 1). Codex is an authoritative body in the interpretation of SPS. SPS also encourages WTO members to harmonize their regulations on 'as wide a basis as possible, [and] Members are encouraged to base their measures on international standards, guidelines and recommendations' (WTO, 2012b: SPS Measures, Art. 3). Codex and ISO are recognized by the WTO and NAFTA as international standards-setting organizations. Members of NAFTA (Chap. 7, Chap. 9, Art. 905) and the WTO are advised to harmonize their standards pertaining to food and agriculture with those set by Codex and ISO. In NAFTA, Codex standards are cited as basic requirements that all members must meet. Since both Codex and ISO's standards operate on the principle of National Treatment, very few PPMs (unless they can be scientifically proven to threaten human, animal or plant health) can be used as grounds to deny the entry of a foreign good into a domestic market or to apply anti-dumping or countervailing duties.

Codex was set up by an FAO/WHO joint initiative to agree on international standards for 'healthy food' in 1961 (Atkins and Bowler, 2001:182). Since Codex's mandate is primarily food safety, its goal is to assure compliance to food standards that apply to the final product (Doyran, 2003:30). The main objective of Codex is to protect consumer health and promote the international trade of food through the harmonization of food standards. It creates fair practices, standards and guidelines for global trade in animal and food products. Codex is recognized by the WTO as the scientific authority on which SPS are based. As stated in SPS text, Codex 'is recognized as the authority for all matters related to international food safety evaluation and harmonization' (OECD, 2003a:133; WTO, 2012b: SPS Art. 3, para. 4).

First proposed by Australia in 1991, FAO and Codex set out to develop a standardized set of international guidelines for organic production. Coinciding with the market for organic products, in 1991 the Codex Committee on Food Labelling considered 'voluntary and mandatory information provision for process attributes' of organic produce, and elaborated on the guidelines for the production, processing, labeling and marketing of organic products (Caswell, 1997:18; Kilcher et al., 2004:28). One of Codex's primary responsibilities is to monitor national organic standards to ensure that they are not acting as trade barriers to other states' organic products (Lohr and Krissoff, 2000:211, Jacobsen,

2002:10). The *ad hoc* Working Group on the Guidelines for the Production, Processing, Labelling and Marketing of Organically Produced Foods met continuously to review the draft guidelines until their finalization in 1999. The Working Group in 1994 included delegates from Australia, Canada, Denmark, France, Germany, Japan, Lithuania, the Netherlands, the US, the EU, IFOAM and the International Organization of Consumer Unions. The group met several times and reported back to FAO and Codex, requesting certain revisions of the proposed guidelines. Issues under review included consistency between definitions and other Codex definitions, livestock production, and labeling provisions for conversion, mixed products (organic and non-organic ingredients) and imports (Codex, 1999: Point 29). Several of the recommendations were to include explicit distinctions around the exclusion of biotechnology from any production system or labeling system referencing organic. In the 1999 report (ALINORM 99/22A), several members of the ad hoc working group wanted to 'include an additional statement to the effect that consumer preferences and other legitimate factors such as ethical concerns should be taken into account in the process' (Codex, 1999:4). This request was not included in the final guidelines. In the document, it is not explicitly stated why. In 1999, FAO and Codex had a series of meetings to discuss the definition of organic agriculture. Together, the two agencies participated in the creation of Document COAG/99/9 that represents FAO's and Codex's positions on organic agriculture. The document defines organic agriculture according to the draft Codex 'Guidelines for labelling'. The final version of the Guidelines for the Production, Processing, Marketing and Labeling of Organically Produced Foods in 1999 outlined more inclusive guidelines for the production, process, labeling and marketing of organic products than had existed previously (Hobbs, 2001b:278; Codex, 2001; Vossenaar, 2003:14). FAO views the guidelines in the following way:

> the standards clearly define the nature of organic food production and prevent claims that could mislead consumers about the quality of the product or the way it is produced. The final objective is to provide the consumer with a choice while giving assurances that organic agricultural standards have been met. (Scialabba, 1999)

It recognizes Codex's role as the standards-setting agency for food products.

The text of the guidelines reflects the broader goals of TBT and SPS as Codex consulted a number of international organizations as well as others representing industry, trade and consumers who share similar goals of

freer trade in organic products (Doyran, 2003:31). Countries like Canada
and Japan have adopted many of the guidelines that Codex has included
in its policy document on organic agriculture into their own national
organic standards. In the event where a WTO member state suspects
national organic regulations are functioning as illiberal barriers to trade,
Codex is used by the WTO in settling trade disputes: 'the WTO may rule
against the importing country if the exporting country is found to comply
with international standards for organic food products, such as those
being formulated by Codex, even if the exporting country does not
comply with the more stringent requirements of the importing country'
(Jacobsen, 2002:11).

ISO

Another organization recognized by the WTO as an authority on international standards settings is ISO. Originating from the International
Electrotechnical Commission, ISO was established in 1947 to create
international coordination and unification of industrial standards. ISO
was created as part of the post-war Bretton Woods institutions along with
the World Bank and International Monetary Fund (IMF). It differs in
terms of its membership, as ISO is decentralized and integrates 135
national standards-setting bodies alongside 2867 technical bodies admin-
istering 12 534 standards. It is a well-established organization that has
been continuously in use since its inception. In the interests of transnational harmonization of standards, many countries like Canada and the
US will voluntarily adhere to the principles of ISO and its standards
when designing national policies. Using well-established international
standards also helps products with developing global markets to be more
easily integrated into existing markets structures, since the rules of
engagement are well known, accepted and established. Developing new
standards-setting organizations or means of certification may require
extensive time and resources to integrate into broader economic systems,
and still may not be generally accepted as equivalent to standards set by
prevailing organizations. Therefore, using standards that are already
widely accepted when designing policies is logical when developing
standards that will be coordinated with pre-existing standards used in
global trade. Despite the voluntary nature of ISO-generated international
standards, ISO has a 'strategic partnership with the WTO' to promote a
free and fair global trading system. TBT as part of the GATT includes
ISO's Code of Good Practice for the Preparation and Adoption and
Application of Standards. Under the TBT provisions, countries file
standards plans with ISO. This creates a standards database and an

architecture that supports a system of coordinated international standards enforcement (Mutersbaugh, 2002:1172). Where international standards already exist, 'the Code states that standardizing bodies should use them as a basis for standards they develop' (ISO, 2006). Much like Codex, ISO's purpose is to harmonize standards internationally to 'contribute to making the development, manufacturing and supply of products and services more efficient, safer and cleaner. They make trade between countries easier and fairer. They provide governments with a technical base for health, safety and environmental legislation' (ISO, 2006).

To adhere to TBT, Canada and the US have both used ISO (also holding official observer status with the SPS Committee) guidelines in fashioning their organic standards. Although diverse national standards for organic agriculture exist, on a fundamental level they must conform to agreed-upon standards and principles embedded in WTO agreements (OECD, 2003b:133, 140). Both the Canadian and American organic programs and regulations conform to international standards regarding certification by complying with ISO's mandate of international harmonization of standards (Bostrom and Klintman, 2006:174; OECD, 2003b:124; USDA, 1999; AAFC, 2014). Both Canadian and American organic certification programs conform to ISO's Guide 65 (general requirements for bodies operating product certification systems), which establishes the generic principles for certification bodies as adapted for organic accreditation by IFOAM (Lohr and Krissoff, 2000:211). Guide 65:1996 was revised to ISO/IEC 17065:2012 in 2012. The overall aim of ISO/IEC 17065:2012 is to certify products, processes and services to provide assurance that each of these meets specific requirements. As stated in the revised guide, 'certification is the degree of confidence and trust that is established by an impartial and competent demonstration of fulfillment of specified requirements by a third party' (ISO, 2012). Adherence to the principles of the updated guide is meant to assure not only consumers, but also clients of the certification bodies and governmental authorities that certification bodies uphold particular standards.

As part of the EU's series of food safety regulations, EU 2092/91 included reference to the earlier version of ISO/IEC 17065:2012 (ISO Guide 65:1996) as the standard to use for organic certification in 1993. In 1999, the EU requested trading partners such as Canada to adhere to the standards it set in 1993 to reduce policy overlap, and ease the movement of certified organic products. As described in an IFOAM–EU document outlining the history of certification: 'imports from outside the EU constituted a significant part of the market and these had to conform to the regulation [EU 2092/91] as well. Thus, the regulation became the benchmark for organic farming around the world' (IFOAM, 2007:9).

With that request came the demand for trading partners to adhere to ISO Guide 65:1996 standards for accreditation and certification embodied in the EU regulation. Transcripts of meetings of international organizations like IFOAM and Codex regarding how ISO Guide 65:1996 was decided upon as the standard template for certification and accreditation of organic products are difficult to trace, but based on interviews completed by Mutersbaugh in 2002, the inclusion of ISO Guide 65:1996 in IFOAM accreditation standards was largely because ISO infrastructure was already well established across many countries, before discussions around international organic standards were underway in the early 1990s. The inclusion of ISO Guide 65:1996 in the first EU regulation regarding organic production further validated the use of ISO standards. In 2003, IFOAM released a document that compared the contents of ISO Guide 65:1996 and IFOAM Accreditation Criteria (Commins, 2003a). In the introduction to the criteria it states, 'the criteria have been strictly based upon the requirements in ISO/IEC Guide 65:1996(E)'. However, organic certification is guarantee of a process and not a product, and this has required some policy adaptation. Despite an overall adherence to the guidelines set out in ISO Guide 65:1996, IFOAM had to modify aspects of the guide to suit the realities of organic production in various environments and countries. But overall, the revised ISO/IEC 17065:2012 is a harmonizing document meant to standardize organic agricultural processes across the board.

Canada's SCC has used ISO standards for decades. Third-party certifiers and provincial certifiers in Canada are accredited under 'ISO/IEC Guide 65' enforced by SCC (AAFC, 2014). Canada and the US also support 'equivalence recognition' as promoted by ISO, which refers to the application of existing regulations by states to ensure that there is some degree of coherence between state policies pertaining to the same issue area. This support for international harmonization of standards was formally recognized with the signing of the equivalency arrangement between Canada and the US in 2009, and similar themed agreements between Canada and the EU in 2011 and the US and EU ratified in 2012. Thus, ISO is an internationally recognized standards-setting organization that aids in the harmonization of certification standards across the globe (Jacobsen, 2002:12). ISO deems IFOAM as the primary international organic standards-setting body.

IFOAM

Although the management of the production processes associated with organic agriculture was traditionally based on a system of self-regulation, organizations at the supranational level have played an important role in guiding organic practitioners as they develop standards and certification requirements based on local conditions. Despite lacking enforcement capabilities, many international organizations developed standards for organic agriculture that preceded national efforts to formally institution-alize organic agriculture into public policy. IFOAM is one such organ-ization. It was established in Bonn, Germany in 1972 responding to the growing interest in organic agriculture in Europe (Willer and Yuseffi, 2004). IFOAM is an independent organic standards-setting organization recognized by Codex and ISO. It also congresses with Codex, ISO and the WTO where organic agriculture is concerned. Its policies are not enforcable, but its positions on particular issues like global trade are considered. IFOAM has 700 members comprised of researchers, certi-fiers, educators and growers who determine the standards that guide the international trade of organic products. It has over 800 affiliates across more than a hundred countries (IFOAM, 2014). As the global market continued to expand for organic products in the 2000s, so too did the prominence of IFOAM in policy discussions concerning organic certifi-cation, labeling, standards and the role of organic principles in organic agricultural practices.

IFOAM first produced standards for organic agriculture called the International Basic Standards for Organic Production and Processing (IBS), in 1980. IFOAM's IBS has three major functions: protecting the organic guarantee from 'field to table', facilitating trade harmonization and avoiding duplication in regulations (Vaupel and Rundgren, 2003:96). IBS sets standards for how organic products are produced, processed and handled that apply internationally. IBS is a model for other certifying agencies, governments and policy makers to use in developing their own standards; however, they cannot be used on their own as a certifying standard (Kilcher et al., 2004:27). IFOAM's standards are used as a benchmark for national organic standards and, as a number of IFOAM documents stress, primarily function to assure consumer confidence in organic labels and products (Westermayer and Greier (eds), 2003). ISO regards IFOAM's IBS and IFOAM's Criteria for Programmes Certifying Organic Agriculture and Processing documents as *the* international stand-ards for organic agriculture – even though the standards set by IFOAM are not legally enforceable (Commins, 2003b:78). This is one of the reasons why IFOAM has linked its standards with other international organizations that have a degree of enforceability to their regulations such as the WTO with its dispute settlement mechanism, and trade-related Codex food safety standards and ISO certification standards.

IFOAM includes a number of social and ecological goals in its mandate which became a clear component of its agenda in the 2000s (IFOAM, 2005b). For example, IFOAM not only promotes and supports organic agriculture worldwide, but it also has taken a stand against the infringement of intellectual property rights on the property rights of farmers and their economic independence (IFOAM, 1999). IFOAM promotes the inclusion of social standards in public policies for organic agriculture and the movement of organic products as outlined in the IBS. As the vice president of IFOAM stated in 2002, 'social justice is part of the organic philosophy' (IFOAM, 2002a). In the IBS' Principal Aims of Organic Production and Processing, IFOAM lists the recognition of 'social and ecological impacts of organic production and processing' and claims to support supply chains that are 'socially just and ecologically responsible' (FAO, 2010). IFOAM also recognizes women's contributions to organic agriculture, and its initiatives include encouraging more women to become involved in organic agriculture and the bureaucratic aspects of IFOAM (FAO, 2010). Labor practices are also a core focus as, according to Barrett et al. (2002:308), 'IFOAM has aims that relate to workers' rights, their basic needs, adequate economic return and satisfaction from their work and a safe working environment. They are also committed to promoting farm organizations to function along democratic lines and uphold principles of equality and power.' IFOAM's interest in labor issues in organic agriculture has led to efforts to link the goals of organic farming on a global scale with those in the fair trade movement. IFOAM's board has devised a project entitled 'Social Audits in Agriculture: developing best practice and co-ordinating with environmental certification in co-ordination with Fair Trade Labelling Organization and Social Accountability International' (IFOAM, 2005c). IFOAM has suggested that the ICFTU/ITS Basic Code of Labour Practice (SA8000 code is a standard that addresses social practices in the workplace) be used in certification schemes to protect labor in organic supply chains (IFOAM, 2004). This particular code addresses issues of child labor, forced labor, discrimination, freedom of association and the right to collective bargaining.

IFOAM, along with the Fair Trade Labelling Organization (FLO), is a member of the International Social and Environmental Accreditation and Labelling (ISEAL) Alliance. This alliance consists of a number of certifying organizations that are concerned with social and environmental criteria in certification schemes. The ISEAL Alliance was formed to gain international recognition for members' respective programmes (Mallet, 2003:89). The stated goal of the ISEAL Alliance is to foster 'positive social and environmental change that ensures a healthier environment and better social and economic conditions for producers and their communities'. Yet the initiatives of ISEAL Alliance are limited to conforming to the criteria based on TBT (Mallet, 2003: 90–91). Standards devised by ISEAL Alliance must not act as technical barriers to trade, and this includes not using PPMs as grounds to discriminate against importing a WTO member's goods.

Importantly, however, IFOAM stresses the issue of equivalence, and promotes the harmonization of national regulations to a basic set of international standards for all organic production and processing (van Elzakker, 2003:82; IFOAM, 2014). IFOAM favors international standards and believes all national regulations pertaining to organic food should harmonize towards the international standards that it and Codex have set (IFOAM, 2005a). It supports the idea that the WTO should not 'hinder non-governmental labeling systems based on PPMs, or other non-product related labeling standards, like organic labeling, environmental labeling or fair trade labeling'. Further, the statement also lends support to the principles of TBT to encourage the use of international standards, such as IFOAM's Basic Standards and IFOAM Accreditation Criteria for Organic Certification (IFOAM, 1999). In addition to Codex's recognition of IFOAM's status in setting organic standards, the WTO and the Organisation for Economic Co-operation and Development (OECD) also recognize IFOAM's position. IFOAM was originally established to harmonize standards developed by private and voluntary sector bodies, and works with other institutions in order to increase the representation of its membership to assure that all certifiers are treated equally and that their policies harmonize with other agencies, such as FAO and ISO. Supporters of harmonized regulation and certification suggest they offer a much-needed degree of confidence to purchasers and buyers of organic products. Proponents of streamlined legislation cite a reduction of certification costs passed on to the customer as a major benefit of policy harmonization regarding the production and trade of organic products (Lohr and Krissoff, 2000:212; Allen and Kovach, 2000:223).

A multi-organization effort to harmonize organic food and agriculture standards in the early 2000s focused on trade in conventional and organic foods. The UNCTAD/FAO/IFOAM International Task Force on Harmonisation and Equivalence in Organic Agriculture has periodically met since 2003 to harmonize international standards regarding organic food and to foster further international trade in organic products by setting regulatory benchmarks while establishing a universal set of standards that protects the integrity of certified organic products (Westermayer and Geier (eds), 2003). Harmonization and equivalence arrangements are argued to be useful in reducing the duplication of policies and certification costs for those involved in the production process. Varying levels of national organic standards and certification and labeling criteria have been recognized by UNCTAD, FAO and IFOAM as barriers to trade in organic products and as a source of consumer confusion as stated in an IFOAM document produced by the task force entitled 'The Organic Guarantee System: The Need and Strategy for Harmonisation and Equivalence' (Rundgren, 2003:6; Daugbjerg, 2012). From 2003 to 2008, the task force focused on various aspects it determined to be barriers to trade in organic agri-food. It conducted studies, proposals and developed tools to address these issues. Two notable tools were developed and published in 2009. The first is The International Requirements for Organic Certification Bodies, which is a document laying out international norms of certification which enable stake holders involved in certification to recognize and accept certification occurring outside of their certification framework. The second is The Guide for Assessing Equivalence of Organic Standards and Technical Regulations, a tool that can help identify equivalence between procedures and criteria between two regions (United Nations Committee on Trade and Development (UNCTAD) et al., 2009:iii).

The next step after the task force completed its mandate is a project entitled Global Organic Market Access (GOMA). GOMA is a platform meant to assist stake holders in implementing the findings of the task force and utilizing the tools it created. GOMA's focus is to work towards international harmonization and equivalence of organic standards and certification to simplify and ease the movement of organic goods around the globe. GOMA tracks various harmonization schemes and equivalency arrangements among trading partners (GOMA, 2012). Though both countries have signed onto various equivalence arrangements, Canada and the US are not part of any harmonization agreements. Harmonized global public policy remains somewhat elusive as two of the largest producers and consumers of organics – the US and EU[3] – have yet to harmonize their standards and certification with other countries.

Nevertheless, these developments demonstrate the significant ways IFOAM's role continues to increase. However, IFOAM has altered its course somewhat, in order to remain relevant in global policy discussions regarding organic agriculture, and must harmonize its regulations and standards to meet those of other WTO-associated organizations like ISO

---

[3]   The European Union has a harmonization regulatory framework in place for organic production that includes all EU countries (EC Council Registration 834/2007). Implementation rules are covered in EC Commission Regulation 889/2008. See EC Agriculture and Rural Development, 2012c.

and Codex. It continues to reject the role of biotechnology in agriculture and subsidy-based agriculture policy that excludes organic production methods (IFOAM, 1999). Initially, IFOAM was created from the idea that the spread of organic agriculture around the world would be beneficial for society because the process-based definition of organic (which it initially followed) promoted social justice, land stewardship and localized markets for organic foods by valuing social, economic and environmental goods throughout the production process (IFOAM, 2002b). Including social principles in enforceable legislation has proven to be difficult as IFOAM harmonizes its standards with WTO principles and agreements that do not include a firm commitment to fair and just treatment of labor or the environmental impacts of production processes as issues that are allowable in policies regarding production regulations (IFOAM, 1999). As one critic of the declining attention paid to labor practices in organic agriculture notes, 'there has been little attempt to make these [labor] codes a legal requirement, not least because to do so could lead to challenges at the World Trade Organization' (Blowfield, 2001:2). Although IFOAM officially holds the position that it supports social justice as an important component of what makes a good organic, it supports policy harmonization with the WTO equivalent with Codex and the global expansion of organic value chains. It thus faces the paradox of promoting the process-based definition of organic which includes elements of social justice, as it facilitates transnational organic value chains that are premised on limiting the inclusion of labor standards and land rights of farmers in standards and certification within a trading system built upon liberal market principles. As Raynolds contends, IFOAM 'embodies sharp contradictions between its original movement-oriented and [its] more recent market-oriented organic norms and practices' (2004:729). According to DeLind (2000:9):

> organic has little hope of succeeding in any meaningful way if its definition is not also predicated on putting more people back on the land, creating useful work that produced a just income collectively in the interest of their own long-term development. Organic without a social vision is dangerously incomplete.

Many of the social and environmental goals championed by the process-based definition of organic are more difficult to realize as organic products enter the global trade system that centres around the concept of product equivalence and denies many issues of process from being included in enforceable standards and certification schemes.

CONCLUSION

This chapter examined how trade agreements and institutions premised on the principles of liberal trade like NAFTA, the agreements enforced by the WTO, and the global organizations that represent interests in expanding trade and markets, present challenges to the ability of supporters of the process-based definition of organic to put into practice production methods that cultivate and value social and environmental sustainability. The product-based definition of organic holds a privileged position in standards and accrediation schemes that are fundamental components to the global trade in organic food. It has also shown that the authoritative and privileged positions of Codex and ISO in interpreting what constitutes 'safety' and 'risk to health' as they relate to food products and production processes have contributed to the institutionalization of the product-based definition of organic at the international level. Despite various efforts by IFOAM to formally promote the substantive aspects associated with the process-based definition of organic, it is unable to withstand the pressure to accept the liberal approach to trade and other efforts to globally harmonize organic standards and establish equivalency arrangements.

Both NAFTA and the WTO are premised on liberal market principles that place emphasis on evaluating the qualities of the end material product, and highly regulate which processes and production methods can be used to erect legitimate trade barriers. They work against including many substantive production processes as enforceable determinants of the equivalence of traded products, which has traditionally been the major difference between organic and conventional agriculture. Considering trade agreements largely represent the product-based definition of organic favored by transnational corporations eager to have markets expand for their products, it is not surprising that the fundamental aspects that separate the product-based and the process-based definition of organic – the 'how' and the 'who' – are absent from enforceable legislation pertaining to production standards for organic food.

6. The development and transformation of the organic social movement

INTRODUCTION

Today organic food is carried in conventional grocery stores and revered in mainstream lifestyle publications as a healthier and more environmentally conscious choice compared to non-organic fare. The origins of organic agriculture in Canada and the US, however, are anything but conventional or mainstream. Beginning in the 1960s, a social movement linked with agricultural practices emerged that began to take on broader social and political goals, many of which came to constitute an organized effort to link the process-based definition of organic with organic agricultural practices. But as the 'organic movement' evolved in response to the popularity of organic food, the movement began to attract a more diverse group of actors with a wide range of interests and levels of commitment to the principles included in the process-based definition. Many of the newer actors to join the movement in the 1980s wanted to capitalize on the market growth organic food was enjoying while engaging with national and sub-national governments to regulate and label organic production processes – two actions the earlier organic movement largely rejected. Some observers viewed the organic movement as a critique of 'productivist agribusiness ... [that] proposed a new vision of society-nature as a whole' (Vos, 2000:251). Though some members of the movement continued to challenge the exploitive relations of the conventional agri-food system by putting the principles associated with process into practice, the membership, objectives and organization of the movement underwent significant change from the 1960s onwards as the growth and influence of other interests transformed the politics of organic food.

This chapter presents the historical development and the transformation of the organic movement. Using the policy process model as an organizing framework, it begins by discussing three elements that experienced significant change in the movement's early formation: the political opportunities, mobilizing structures and framing processes. The second section presents the transition of the organic social movement into what is defined as a 'new social movement', focused on the public promotion of marketable qualities of organic food products rather than the social and political objectives of the earlier movement, which essentially undermined any organized attempt to firmly link the process-based definition with organic agricultural practices. The organic movement was vulnerable to the shift in focus for two central reasons. The first is the over-reliance on the market as the primary mobilizing structure, and the second is the lack of formal controls over membership in the movement. These factors allowed actors to enter the sector that prioritized market expansion over transformative socio-economic goals that defines social movements as distinct from other forms of activism.

THE ORIGINS OF THE ORGANIC MOVEMENT

The organic movement in Canada and the US largely remained a series of disaggregated groups of practitioners and supporters of organic agriculture until the countercultural movements of the 1960s added a distinctly 'social' element and a clear set of 'political' goals. The early organic movement was part of the critical 'bottom-up' element of civil society whose members collectively challenged the dominant industrialized agri-food system (Cox, 1999:7). Conceptualizing the early organic movement as part of the ongoing process of resistance to the dominance of industrialized agricultural practices can help explain the changing politics of organic food and the changing orientation of organic advocates and practitioners towards the state and the market (Sumner, 2005).

Modern social movements concerning food and agriculture have a long history, beginning with the agrarian peasant movements of the early to mid-twentieth century. These movements were based on regaining the economic security of agrarian traditions from market-led transformation (Paige, 1975; Freyfogle, 2001). Some early organic practitioners were gardeners and farmers, primarily interested in learning and practicing organic techniques such as composting to regenerate soil health. But others believed that the holistic approach of organic agriculture could be used as a blueprint to mount a social challenge to inequalities found in modern society, namely those found in conventional agriculture. Those who attached political significance to organic agriculture's ideological opposition to industrialized forms of agriculture looked to early agrarian movements that advocated working with nature and the soil as prescribed by Steiner, Howard and Balfour (Peters, 1979). The most influential force that helped to transform the organic movement's primary focus on soil health was the emergence of 1960s counterculture in North America.

The 1960s counterculture emerged as a series of social movements meant to combat the social, economic and political inequalities in advanced capitalist democracies. Social movements, according to O'Brien et al. (2000:12), are 'a subset of the numerous actors operating in the realm of civil society. They are groups of people with a common interest who band together to pursue a far-reaching transformation of society. Their power lies in popular mobilization to influence the holds of political and economic power'. This definition captures the nature of the early organic movement – members of society had the vision of reforming agricultural practices by presenting a viable alternative to the status quo. Bostrom and Klintman (2006:167) further clarify the organic movement as 'cases where the actors are fully devoted to the idea, principles, and practices of organic production, as a cultural challenge to conventional/industrially oriented agriculture'. Referring to social movement literature is useful in assessing how the early organic movement in Canada and the US developed over time. Studies of social movements focus on how they are established, evolve and meet their goals. The policy process model used in examining social movements looks at three factors that experience change as movements mature: political opportunities, mobilizing structures and framing processes (Tilly, 1978, 2004; Tarrow, 1983, 2005).

Political Opportunities

Political opportunities are defined as the political climate of a society that sets the boundaries and possibilities for the establishment of a social movement. Political opportunities address how political constraints impact the development of a social movement and the opportunities for collective action, as well as the ability of a social movement to meet its desired outcomes, what are labeled 'goals and objectives' (McAdam, 1982). Economic constraints factor into the availability of political opportunities as well. In many cases, it is the political institutional setting which experiences some sort of change that stimulates the establishment of a social movement. In the case of the early group of organic practitioners, there were two distinct political opportunities that arrived in the late 1960s that helped to transform it into an identifiable social movement. The first was the growing public awareness of the negative outcomes of industrializing processes of the economy. The second change was the existence of social movements that were extremely critical of the socially and environmentally exploitive nature of industrial forms of production. Industrialized agriculture was well established in both Canada and the US in the late 1960s, leaving little room for the inclusion of organic principles associated with the process-based definition of organic in agricultural policy. The growing public awareness of the negative social and environmental outcomes associated with the rise of the industrialized model of agri-food production encouraged certain segments of society to question the capital accumulation of agribusinesses and their cooperative relationships with the state. The early group of organic practitioners began to promote a set of ideas that represented their interests in expanding organic agriculture, and used the environmental and social degradation endemic to industrialized agriculture to transmit their message to the public. At the time, the general population was also convinced by the scientific community of the superiority of industrialized forms of farming in terms of productivity (Bookchin, 1976; Merrill (ed.), 1976). Because of the growing awareness of environmental abuses and the exploitation of labor found in industrial models of agri-food production, supporters of organic agriculture began to mount a political challenge to the status quo.

The expanding industrialization of agriculture was an important political opportunity for the creation of the organic movement. But because organic agriculture had been largely rejected by the state and mainstream market actors since World War II, another political opportunity was required in the late 1960s that would help to mobilize a greater number of people into action to support organic agriculture in a collective way. The political and economic climate of the 1960s, including the Vietnam War and the coming of age of the 'baby boom' generation, set the stage for some form of social protest to emerge (Dalton and Kuechler (eds.), 1990; Tilly, 2004). The middle class established in the post-war era allowed many young people to attend post-secondary institutions that encouraged them to think critically about mainstream society's social norms and values. Having increased access to information through advancing telecommunications networks also helped to educate young people about the inequalities occurring in the rest of the world and within their own borders. People in industrialized societies began to question the legitimacy of various spheres of power (military, political, economic) and the role of power in creating these inequalities. A number of social movements challenging the norms and values of mainstream society began to emerge as a result. The anti-war, civil rights, women's and environmental movements all gained support from the middle class youth who were dissatisfied with the status quo and motivated enough to act upon these convictions (Belasco, 1989).

As mentioned in Chapter 2, the feminist movement of the 1960s brought with it principles of gender equality that appealed to some organic practitioners and supporters who rejected conventional social relations, but it also raised important issues surrounding social equality and power hierarchies in general. Food issues were important to women, as women traditionally are the primary purchasers and preparers of food for themselves and their families. Many feminists at the time, who were concerned about food safety and therefore interested in sourcing healthy and cost-effective food for their families, were eager to get involved in a type of food production that deviated from practices in conventional agriculture that caused harm to evironmental and human health.

For women who did not have access to credit to purchase land, or expensive machinery, organic farming was a cost-effective way to grow a variety of food on small plots of land (Allen and Sachs, 1991; Chiappe and Flora, 1998). Norwegian researcher Hilda Bjorkhaug links the principles of the feminist movement and the organic movement, noting that 'the "organic" ideology has several links to what might be called the feminine principle' (Bjorkhaug, 2004:5). The strongest link between the feminine principle and the organic approach to food production is the shared notion that processes need to be recognized as integral parts of outcomes. The feminist movement and its recognition of process added a distinctive social context to the emerging organic movement, as other social movements of the era such as the sustainable agriculture movement did not address issues of gender equality (Allen and Sachs, 1991; Mearnes, 1997).

In addition to the feminist movement, the sustainable agriculture movement helped to establish some principles that the organic movement would also promote, though each movement had differing world views. The sustainable agriculture movement promotes the idea that agriculture should be practiced using ecological principles of achieving food self-reliance, a concept of land stewardship and the sustainability of rural communities. The movement was critical of the use of technology in industrialized agriculture that displaced rural farming communities from the land and supplied the general population with highly processed and nutritionally deficient food (Henderson, 1998:113). By vilifying agribusiness as the enemy of family farmers and the concentration of land ownership as a major threat to the agrarian traditions of rural America, the populist goals of the sustainable agriculture movement were able to highlight the exploitive nature of mainstream, industrial agriculture (Youngberg and Buttel, 1984:174). Attempts were made by members of the early sustainable agriculture movement to mobilize the general public by appealing to those who objected to capital accumulation by agri-business and rural displacement.

To present an alternative to industrialized agriculture, the sustainable agriculture movement promoted the idea that agri-food production should return to its agrarian roots. This entails reconnecting people to the land through localized market interactions, such that consumers were directly connected to the producers (Schumacher, 1973; Berry, 1977). The movement was not interested in lobbying the state for agriculture policy reform. 'Back to Landers' in the late 1960s rejected engagement with the state entirely and had no interest in influencing or changing political institutions. They were more interested in maintaining a distance between their private activities and what they viewed as the intrusion of govern-ment, thus reflecting a libertarian political view. Wendell Berry's *The Unsettling of America* (1977) spoke to supporters of the sustainable agriculture movement who were concerned with the restoration of the pastoral landscape in the US. Visions of this landscape presented by Berry in his writings served as important motivational images for members of the sustainable agriculture movement to participate in re-establishing the disappearing family farm and practicing self-sufficient farming (Belasco, 1989:76). Berry's ideas transcended borders and his works were equally influential to Canadians who were concerned with the rapid depopulation of the rural countryside.

The sustainable agriculture movement offered the emerging organic movement a number of ideas to internalize in practitioners' advocacy for organic agriculture, but in some ways they remained different. While there are clear ideological links between the environmental movement and organic agriculture, there is no consensus as to the strength of the links. Some scholars investigating the socio-politics of agriculture, such as Buttel, argue that reforming industrial agricultural practices was not a central element to the platform of the mainstream environmentalist movement and that the issue linkage between agricultural and environ-mental movements is often overstated (Buttel, 1997:358). Gilman (1990:10) notes that often the goals of the sustainable agriculture and the environmental movements are conflated, pointing out the primary differ-ences between the two:

> [The sustainable agriculture movement] *insists* on a whole-systems approach, whereas the environmental movement has focused on the human impact upon non-human systems ... Unlike much of the environmental movement, it is vision and solution-oriented ... it is primarily concerned with the nuts and bolts of ecological and cultural health.

However, in their early forms, both the environmental and sustainable agriculture movements challenged the status quo of the 1960s. They were part of the counterculture that asked important questions about agriculture, laying important ideological foundations for the organic social movement to be established.

The issue of food quality and its relationship to environmental health forged an important link between the sustainable agriculture, environmental and organic movements (Levenstein, 1993:184). The rise in general awareness of environmental problems, and the dangers associated with chemical use due to the popularity of several publications, helped to draw attention to the troubles of the industrialized agri-food system (Marshall, 1974:51). Rachel Carson's book *Silent Spring* (1962) is often identified as an important publication that helped to bring awareness to the public of environmentally damaging practices. Environmentalists also had concerns regarding the chemicals used in industrialized agriculture and the environmental implications of their use. Meadows et al.'s *Limits to Growth* (1972) had an apocalyptic Malthusian vision of the future because of industrial economies' dependence on finite natural resources and global overpopulation. This book is often referred to as an important driver in establishing the environmental movement as it drew attention to the environmental and social risks associated with heavy reliance on non-renewable resources. Levenstein's chronicling of eating habits in the US attributes the rising public interest in organic food in the 1970s to the growing fears of DDT and pollution, issues commonly associated with the environmental movement (Levenstein, 1993:162). The environmental movement rejected the production processes involved in over-processed food and encouraged people to move towards a 'whole food' vegetarian diet that relied less on fossil fuels and food processing, as prescribed by Frances Moore Lappe's *Diet for A Small Planet* (1971). This added another link between environmentalism, sustainable food production and organic agriculture.

Both the environmental and the sustainable agriculture movements presented a challenge to the status quo and the industrialization of agriculture. They both rejected the industrializing processes of modern society but they were quite different in terms of their goals and objectives. Early environmentalists campaigned for less pollution and environmental degradation, while those in the sustainable agriculture movement had the goal of getting people back on the land. The sustainable agriculture movement did not take issue with whether or not farmers practiced chemical agriculture. Neither movement fully covered the aims of the developing organic movement, which included critiques of social, economic and environmental dimensions of conventional industrial agriculture. The 'Back to the Land' movement was more concerned with repopulating rural landscapes without directly challenging the wider social ills associated with the industrialization of agriculture (Jacob, 1997:5). In social movement literature, the types of social movements promoting a return to simpler ways of living (less reliant on highly manufactured goods) have been described as 'anti-modern' movements that idealized the pre-industrial styles of living as the 'source of moral and physical recovery' (Brand, 1990:29). In contrast, the organic movement promoted organic agriculture for a modern society.

Similarly, some supporters of organic agriculture did not want to associate themselves with the environmental movement of the 1960s because they felt it did not address many of the social issues linked with sustainability. As radical ecologist and libertarian Bookchin (1976:10) observed:

> [I]t may be well to distinguish the ecological outlook of radical agriculture from the crude 'environmentalism' that is currently so widespread. Environmentalism sees the natural world merely as a habitat that must be engineered with minimal pollution to suit society's 'needs', however irrational or synthetic these needs may be. A truly ecological outlook, by contrast, sees the biotic world as a holistic unity of which humanity is part.

Others saw a global crisis looming in the environmental movement and viewed the creation of an organic movement as a far more viable option that addressed the social, economic and environmental degradation associated with the industrialized agri-food system by linking it to negative effects on the health of the soil and nutrition content of food. As the organic movement in the 1960s continued to develop, some members began to ask questions about the role corporate actors played in industrialized agriculture and the decline of the 'family farm'. Periodicals such as the *Whole Earth Catalogue* (*WEC*) and *Organic Gardening* (*OG*), though not explicitly providing ways for readers to challenge the social ills associated with industrial agriculture, promoted the idea that participating in organic agriculture as a producer and consumer were ways of socially challenging the status quo. The idea that food production should be practiced in a holistic manner with the environment was a strong message sent by both periodicals. In Berry's *WEC* review of Howard's *Agricultural Testament*, he argued, 'the scientific respectability of organic methods has been obscured for us by those [who] have insisted upon making a cult of the obvious and by the affluence and glamor of technological agriculture – the agriculture of chemicals and corporations' (Berry, 1969:33). In a later issue, Berry continued to emphasize the social and political issues linked with organic agriculture. In his article 'Think Little', Berry (1970:4) wrote:

> [T]he American farmer is harder pressed and harder worked than ever before ... his margin of profit is small, his hours long ... For the small farmers who lived on their farms *cared* about their land ... The corporations and machines that replace them will never be bound to the land by the sense of birthright and continuity and love which enforces care.

Another strong message promoted at the time was the link between organic food and a healthier lifestyle. In an article chronicling the Sourthern California organic movement published in 1970, a contributor to *Organic Gardening and Farming* (*OGF*) (later renamed *Organic Gardening* in 1973 and relaunched as *Organic Living* in 2015) magazine interviewed consumers of organic foods. Goldman (1970:40) wrote:

> they buy organic because of quality and flavor ... people are just waking up to the fact that *all food should be health food*. With the spotlight on pollution, poisons and the ecology in general, it's apparent that even infinitesimal amounts of these deadly poisons on foods are literally time-bombs which may have fuses that can go off anywhere from now to five, ten, even 20 years later.

Periodicals focusing on organic agriculture were important ways of disseminating messages to their readers. *OGF*'s emphasis on organic food as more nutritious and healthier than conventional fare was an important way to attract new readers concerned with the safety and quality of the agri-food system. This expanded the scope of the member-ship including the more radical members that Wendel Berry spoke to, but also to a more general audience in the mainstream society. Though *WEC* did not directly speak to issues of process as understood in the process-based definition of organic such as labor issues or the idea of 'localized' food chains, writers like Berry and those contributing to the *WEC* presented ideas challenging the social and political norms that accom-pany industrialized agriculture. They drew attention to the plight of the 'small-scale family farm' but also to some of the perceived negative aspects of industrialized agriculture such as the use of petroleum-based inputs in food production. But publications like *WEC* focused more on the anti-corporate message than other principles of the process-based definition of organic. The *WEC* provided a platform to address some of the social problems emerging from the consolidation of the agriculture sector in general, but it did not emphasize labor practices, or engage with the democractic system to fight for change. The lack of attention paid to the social dislocation of farming communities that followed the consolidation of agricultural land and resources certainly did not help to elevate the importance of labor issues in the organic sector.

Mobilizing Structures

In addition to assessing political opportunities the policy process model also addresses mobilizing structures of social movements. The concept of a mobilizing structure refers to the organization of a social movement and the networks – both formal and informal – that exist to help mobilize people and engage them in collective action (McAdam et al., 1996:2). In essence, mobilizing structures refer to the *means* social movements use in meeting their objectives, including political, social and economic resources available to mobilize the population. In many cases, mobilization is directed towards lobbying the state and other political institutions to stimulate social change. However, unlike the environmental movement in its lobbying the state to restrict the use and abuse of environmentally hazardous materials and practices, the early organic movement rejected engaging with the state in achieving its goals, in large part because of its anarcho-libertarian ideas borrowed from the sustainable agriculture movement and the 'Back to Landers'. Although not engaging with the state afforded organic practitioners significant independence in determining how social protest would take shape, the disaggregated nature of mobilization also limited the amount of political resources it could draw upon (Berry, 1976; Egri, 1994:150). The organic movement in its most radical form wanted to create an alternative mode of social organization that was independent from state involvement, and practitioners turned towards the market as its major resource in mobilizing the population. The market was viewed as the primary mobilizing structure for the movement.

In its earliest days, the organic movement, like most other social movements in their infancy as shown by Meyer and Tarrow (1998:19), had little in terms of well-structured organizations to mobilize supporters. The best means possible for the widest distribution of the social and political goals of organic agriculture was determined to be through market transactions and consumer choice. Members of the early organic movement believed that if the public had access to the right information, it would choose organic over conventionally produced foods, which would eventually lead to more land used for organic food production, resulting in sector-wide change in agriculture. Since many food corporations in the conventional market rejected organic food's health and environmental claims, the only way to make organic food available to the

public was through alternative markets that did not require the involvement of actors or institutions from the mainstream. Members of the early
organic movement believed that creating alternative food networks was
necessary to circumvent the destructive involvement of the state and
agribusiness in food production, and could in fact change market
relationships themselves. The belief that promoting the consumption of
organic food was the best way to achieve change is reflected in the
comments of a member of the early organic movement: 'we should cling,
with whatever optimism possible, to the idea that the same economic
forces that brought us environmentally bad products will be the ones to
get them out of the marketplace ... the organic force that is surfacing ...
has the elements to revolutionize the marketplace' (Goldstein, 1976:223).

The organic movement looked at the market as a politically neutral
institution that was able to accommodate an alternative form of production and distribution to conventional means. The market was viewed as a
powerful vehicle to distribute organic food to the public and also to
recruit more members into the movement by using interactions between
producers and consumers as the opportunity to educate people about the
social, environmental and economic benefits of organic agriculture. Since
the early movement lacked a cohesive form of organization, informal
grassroots networks were established as membership grew. In the 1960s,
membership in the organic movement expanded to include food cooperatives, small health food stores and local agricultural associations, in
addition to organic farmers and purchasers of organic foods (Guthman,
2004:6).

Those in the organic movement were determined to use food cooperatives as a way of supporting organic producers and organic principles. Making organic food available to the wider public and spreading
information about organic farming through the cooperative market model
was viewed as a way to help mobilize the public against the industrial
agri-food system. The mandate of food cooperatives was to sell foods
that were sourced as locally as possible (Sligh and Christman, 2003:33).
As discussed in an *Ms.* magazine article on women and organic farming,
women were often at the helm of food cooperatives in the 1960s and
1970s, seeking alternative ways to source healthful food products in bulk
quantities (Lipson, 2004). Small-scale organic outlets like farmers'
markets and independently owned health food stores were viewed by
members of the organic movement as 'symbols of the new America
Revolution', and as an alternative to the conventional agri-food system
because they cut out the 'middlemen' from the value chain (Belasco,
1989:73). The logic was that this form of market organization would
ensure that organic producers and supporters of the process-based

definition of organic would retain the value added to organic food products throughout the production process.

Although it is difficult to determine how large the early organic movement was in terms of the number of members, or how successful it was in spreading its message to the public, the rapid growth of alternative food networks gives some indication as to the successful use of the market as a mobilizing structure. It was estimated that in 1965 there were about 500 'health food stores' in the US (Wolnak, 1972:455). Between 1969 and 1979 approximately 5000 to 10 000 new food cooperatives were established across the US (Belasco, 1989:90). In California alone there were over 300 health food stores and 22 organic restaurants in 1970 (*Newsweek*, 1970:100). According to Marshall (1974:50), by 1972 the number of health food stores had increased to over 3000. Even at this early point in the emergence of the organic movement, select supermarkets and department stores began to carry both kinds of 'health' and 'organic' food products. Other estimates, such as those reported by the Organic Gardening and Farming Association, claimed that by 1972 there were approximately 3477 organic food stores in the US (Myers, 1976:136). The number of health food stores in Toronto jumped from 13 in 1957 to over 100 in 1979. The reasons for the market growth in health foods in the 1960s and 1970s in Canada were very similar to what influenced the market expansion in the US, including the influence of Rachel Carson's book and changing public perceptions and awareness about pesticide residues on food (Cooper, 2006).

In 1971, Arran Stephens (who would go on to establish British Columbia-based Nature's Path Foods in 1985) established Canada's largest organic supermarket, LifeStream. The goal of the supermarket chain was to provide consumers with wholesome, healthy foods. With the success of LifeStream, Stephens then expanded the company into food processing including milling grains and making breads and cereals. By 1977, LifeStream was grossing $9 million (CAD), making it one of Canada's most successful natural food retailers (Nature's Path, 2013). It was so successful that LifeStream published a vegetarian cookbook that sold over 125 000 copies in Canada. In both Canada and the US, advocates of organic agriculture shared a common set of values and 'the idea of natural, organic farming became the basis for a consumer movement; health food stores spread, and their products even entered the chains as consumer knowledge and independent buying habits grew' (Myers, 1976:136).

Increasing the consumption of organic food was promoted as the primary means of achieving social change. By purchasing food that was not associated with state-subsidies or agribusiness, consumers could 'vote with their dollars' and protest against the industrialized agri-food system through alternative market transactions. As one supporter of organic agriculture argued in the 1970s:

> when you buy organically grown foods produced by a family farmer who is not supposed to be able to make a living on the land, you become an organic force helping to reverse a trend that has driven people off the land and made farming the profession of an old generation. (Goldstein, 1976:215)

Through conscious consumption via informal networks of organic production, consumers could help to sustain the rural way of life and the overall health of the environment through alternative market transactions that were able to function outside of the mainstream market.

Framing Processes

One way of understanding the motivation behind individuals' support of organic agriculture through their membership in the organic movement is what social movement theorists call 'framing processes'. Framing processes refer to the shaping of the subjective reality of individuals and how this subjective reality influences the shape of social movements (Snow et al., 1986). Essentially, framing processes address how social movements are organized in terms of their membership and how the values and norms of the membership (individuals or associations) influence the movement's direction and overall goals (McAdam et al., 1996:5).

The framing process that helped to attract membership to the organic movement stemmed from the existence of other countercultural movements made up of those who disagreed with the socio-economic inequalities of the status quo and industrialization's effect on the environment. Many participants in social movements of the 1960s were recruited from the educated middle class who questioned the exploitive nature of capitalist forms of production. Movements like the environmental movement gained notoriety not only because the environment was more damaged than it was in the past, but also because the public was more aware and sensitive to the quality of the environment in which they lived (Ingelhart, 1990:44). Individuals who became educated about environmental and agricultural issues were moved to action, critiquing the negative outcomes of the industrial agri-food production system (Friedland, 1994:219). Since links were made between the health of the environment and the nutrition content of food, the organic movement was able to attract a diverse membership including people concerned with the issues that overlapped the process-based approach to organic food production with the environmental movement. A 1970 *Newsweek* article addressing the growing diversity of the organic movement states, 'for years, organic gardeners have been considered part of the harmless lunatic fringe, along with flat-earthers and UFO spotters ... [but] the organic-food community now includes not only ... vegetarians [and] macrobiotics ... but large members of environmental activism, house-wives with tired blood and sophisticated gourmets' (*Newsweek*, 1970:100). As the organic movement gained support from members of the broader public, it successfully gained media attention that helped to change societal perceptions of organic agriculture, the people involved in it, and its benefits beyond nurturing the health of the soil and producing 'healthy' food.

The way in which ideas were transmitted to members of the organic movement and the general public helped to frame the issues so that people were prompted to organize. Publications were crucial for dis-seminating ideas to the general public and encouraging concerned persons to act collectively. A number of publications promoted the objectives of the organic movement in an effort to mobilize readers to act, although not all explicitly discussed or mentioned the association between organic agriculture and social activism. Although J.I. Rodale was not interested in promoting organic agriculture's association with the political radicalism of the counterculture of the 1960s, he did realize that younger people's increased understanding of the problems of industrial agriculture and its associated environmental impacts were essential in expanding organic agriculture (Levenstein, 1993:198). Those who were concerned with social, political and ecological issues in the 1960s recognized organic agriculture as a method of food production that challenged the status quo by remaining politically and economically independent from agribusiness and government. The challenge which organic agriculture presented to the general public helped to fully establish the organic movement as a 'social' movement. *OGF* promoted similar ideas about more natural gardening and farming techniques while *Prevention* magazine oriented itself more towards providing health and nutrition information associated with food products. *OGF* and *Prevention* were important sources of information for the informal, disaggregated membership of the organic movement, with *OGF* focusing its content on organic gardening techniques while *Prevention* included articles on health and nutrition (Raeburn, 1995:226). The WEC was also an import-ant part of the framing processes for the organic movement at this time.

Although the *WEC* served the wider counterculture as well as the organic movement, much of its contents from its inception in 1968 proved to be useful to practitioners of organic techniques who relied upon information about low-tech sustainable agricultural techniques (Armstrong, 1981). As stated on the inside cover of every issue of the *WEC*, its purpose was to 'develop power of the individual to conduct his own education, find his own inspiration, shape his own environment' in response to the 'power and glory – as via government, big business, formal education, church – [that have] succeeded to the point where gross defects obscure actual gains' (*WEC*, 1970). The purpose of reviewing organic gardening books was to inform readers about alternative ways of living, while attaching some political significance to action. Since the *WEC* was pitched to those who felt disenfranchised by their governments and society in general, it was a useful publication in getting the process-based definition of organic out to the socially concerned public, though its contents are by no means representative of the diverse group of people who considered themselves part of the organic movement. Because of the *WEC*'s staunch anti-corporate and anti-government position, it was an ideal publication for spreading ideas associated with the organic movement. Although the *WEC* was only in circulation for four years, it had a lasting impact on recruiting new membership for the movement. As its readership expanded, it served as a global hub in an information network that continues to challenge the status quo. According to the Whole Earth website (which replaced the printed version some years later), the *WEC* was the 'unofficial handbook of the counterculture' (www.wholeearthmag.com).

Another influential publication that helped in issue framing was *Mother Earth News* (*MEN*). *MEN* was pitched to members of the organic movement as it was full of recipes and suggested that readers try various organic and natural foods, while also catering to members of the 'Back to the Land' movement. *MEN*'s advertisement in the *WEC* stated, 'the MOTHER EARTH NEWS is … for today's influential "hip" young adults. The creative people. The doers. The ones who make it all happen. Heavy emphasis is placed on alternative lifestyles, ecology, working with nature and doing more with less' (*WEC*, 1970:19). *MEN* highlighted issues of ecology and self-sufficiency in its progress – two of the qualities associated with the organic movement. *MEN* is now exclusively an electronic publication geared towards those who want to practice self-sufficiency and sustainability. By 1972, the US circulation of *MEN* reached 60 000 and then grew to 600 000 paid subscriptions in 1980 (Armstrong, 1981:197). Its purposes and goals have changed little since they were first introduced in 1970, and much of the information it continues to publish remains relevant to practitioners of organic agriculture.

Feminist publications focusing on organic farming emerged from women's desire to educate each other and to find new and different ways of sourcing and preparing healthy food. Publications emerged that helped to spread information about more healthy and environmentally friendly ways of eating. In 1973, a feminist periodical called *Country Women* was launched in the US as a source of information for women who were interested in farming without expensive chemicals and machinery, learning new recipes and communicating with other women who were equally concerned about food issues (Belasco, 1989:82).

Although the *WEC, MEN* and *Country Women* were US-centred publications, they were also distributed in Canada. Canada developed its own periodicals about health food and nutrition. *alive* magazine, for example, launched in the 1970s and continues to provide Canadians with news and information about health food and healthy living. It departed from other periodicals by primarily focusing on health food over agriculture or gardening techniques. In this sense, *alive* commented on the perceived health and nutritional benefits attributed to organic food, but did not focus its content on agriculture or draw attention to social issues related to conventional agriculture. *Healthful Living Digest* established in the 1940s in Winnipeg offered its readers information on nutrition, alternative forms of health care and organic agriculture. The early emergence and ongoing existence of health-related periodicals in Canada shows that some of the same concerns about the industrialized agri-food system highlighted in US publications were also addressed in Canada (Cooper, 2006; Castairs, 2006). But the periodicals mentioned here focused on the health and nutrition benefits more than the evironmental benefits of organic agriculture. In this sense, there were two streams of media addressing different aspects of organic food; one that included *OGF* and *WEC* that put more emphasis on the processes involved in organic agriculture, and others such as *Prevention* and *alive* that focused their content on health and nutritional aspects of food consumption. As the market for organic food began to change, publications focusing on health and nutrition gained broader appeal as more information became available to the general population regarding health, nutrition and food.

The organic movement emerged largely as a reaction to the political institutions that privileged industrialized modes of production over the more traditional, agrarian principles associated with the process-based definition of organic. The inability for the environmental and sustainable agriculture movements to address all of the social, environmental and economic issues that organic agricultural associations were concerned with also protected space for the establishment of a coherent organic movement with a clear set of goals and objectives for collective action.

The early movement, complete with its goal of revolutionizing agriculture, attracted a solid membership and used the market as its mobilizing structure because it was viewed as a politically neutral institution. However, in the coming decades other factors such as changing political circumstances and demands of consumers would shift the goals of the organic movement away from its origins.

THE ORGANIC MOVEMENT IN TRANSITION

Once the early group of organic practitioners began to resemble a social movement, its goal was to further expand its membership. Paradoxically, its expansion in many ways threatened its identity as a 'social' movement. What was necessary to reach the broader public, was a message that spoke to people who were worried about the environment, and the negative effects of industrialized agriculture on food safety and human health. Even as early as 1971, J.I. Rodale questioned where the organic movement was headed. When interviewed for an *NYT* article he stated, 'we are afraid of becoming legitimate ... I don't know how to operate if we're in a majority' (Greene, 1971:68). While Rodale feared what would happen if the movement catered too much to mainstream interests, others doubted the future of organic agriculture all together, seeing little future for the movement, at least in terms of its mainstream market potential.

Criticisms of the emerging movement made their way into scholarly journals such as *Food Technology*. In 1972, one article argued that the expansion of organic agriculture was ethically unsound, noting that 'if [corporations] were to expand into the organic food market, [they] would be supporting a cause which seems to be based on misunderstanding and fear ... if [corporations] were to enter the organic market, [they] would be endorsing a movement which ... would retard efforts to increase world food supply' (White, 1972:33). In another article entitled 'Health Foods: Natural, Basic and Organic', 'food faddists' were argued to be driving the growth in the health food sector. Organic foods were criticized as 'generally inadequately prepared, have little or no increased nutritional value, are usually poorly packaged and presented, and have poor storage characteristics' (Wolnak, 1972:454). In 1972, the 'Institute of Food Technologists' (IFT) Expert Panel on Food Safety and Nutrition' outlined the differences between organic and conventional foods. Despite popular claims that organic food was more nutritious than conventionally produced food, the committee presented a dissenting view. It stated that the reasons why organic food was gaining popularity was its positive associations with nutrition, but also organic agriculture's practices of composting and using humus, and its integrated pest management strategies. But beyond those differences, the piece highlights the role which perceived 'value' plays in consumer decisions. As the committee's report concludes, 'knowledge of the emotional value of food is much older than the nutritional value of food; hence, many people will pay the increased price because food has an emotional rather than intellectual value' (IFT, 1974). This 'emotional value' and the association of organic food with health and nutrition regardless of evidence to prove or disprove those claims, is what would help propel the organic movement into the mainstream; by focusing on the health value of the organic food itself and the belief that purchasing organic foods was a healthy choice for people and the environment. Though this collection of commentaries on the state of organic food in the 1970s questioned the focus on the nutritional and health claims made by proponents, they show that this was increasingly the way in which organic food was viewed in the mainstream – as part of the health food movement.

The organic movement persevered and expanded in ways somewhat different from what early practitioners like J.I. Rodale imagined. However, it moved beyond its initial radicalism of forging alternative agricultural networks as it became integrated into the mainstream food retail market. The focus of some advocates of organic agriculture shifted from materialist concerns relating to transforming the exploitive economic and political structures of the agri-food system, to appealing to the public's growing concern over food safety and health. In a 1987 *NYT* article commenting on the growth of the organic food sector, secretary of the Organic Food Production Association of North America, Judith Gillan stated, 'the attitude of the 1960's had been mainstreamed … you don't have to be out in the woods with your hair down your back and refuse to talk to people who go to Wall Street every day to care about organically grown food' (Burros, 1987). The promotion of organic food as a safer, healthier and more environmentally friendly option was far more compatible with the 'post-materialist values' of 'new' social movements (NSMs) over the materialist concerns of environmental and social sustainability.

Ingelhart and Appel (1989:45) define post-materialist values as ideals that put a 'greater emphasis on such goals as self expression, quality of life and belonging and are associated with a decline in traditional values'. Materialist values, in contrast, are defined as those that emphasize economic and physical security. Scholars studying social movements attribute the rise of NSMs in the 1960s and 1970s to growing public concern with issues of quality of life within industrialized societies. Because of the post-war prosperity and absence of armed conflict in liberal democratic countries in Europe and North America, according to

Ingelhart and Appel the post-war generation put less emphasis on economic and physical security than previous generations that lived through two world wars and the Depression (Ingelhart and Appel, 1989:46). NSMs moved beyond the goals of previous movements that had sought to answer questions of economic and political power and redistributive issues, such as the labor and agrarian movements (Brand, 1990:25). Scholars have characterized the emergence of NSMs as a reaction to the resultant problems associated with industrialization and technological development, and their negatively perceived social, ecological and economic outcomes (Offe, 1985). Others have identified NSMs as symptoms of a paradigm shift in values and an increasing conflict between old materialist concerns and newer post-materialist preferences (Ingelhart, 1990). NSMs, with their focus on post-materialist values, aim to mobilize ideas and values to stimulate collective action and to achieve social change as opposed to drawing on the material interests of collective identities, such as class.

Many NSMs, such as the environmental movement, aim to transform the world by providing an alternative to the mainstream 'technocratic and bureaucratic socio-political systems' based on material consumption (Ayres, 1998:13). Movements such as these have the goal of making an alternative world vision a reality. The organic movement as it existed in the 1960s and 1970s included the materialist concerns (economic security) of 'old' social movements and post-materialist concerns (quality of life) of 'new' social movements. Since the early organic movement included some objectives of the agrarian movement (land repatriation) but also some from NSMs like the environmental movement (ecological conservation), it contained both materialist and post-materialist goals in its objectives, making it somewhat unique compared to other movements at that time. But as political and economic circumstances changed, its goals became far more focused on post-materialist values such as food quality and safety and the environmental benefits of organic farming, which appealed to mainstream consumers.

The shift away from the 'conformity' and 'ideology' of social movements to gain broader membership bases is what della Porta and Tarrow label 'progressive politics' (2005:3). Class-based politics faded from the mainstream social movements and more ideas-based movements emerged that spoke to the public's concerns over the sustainability of industrialized production processes and their dependence upon non-renewable resources (Hechter, 2004). Hall claims that the rise of values and ideas-based movements and their corresponding networks have been by-products of the post-Fordist decline of industrialized economies and thus class identification among individuals. This has created what he characterizes as 'fragmentation and pluralism ... [a] weakening of older collective solidarities and block identities' (Hall, 1991:58). Because of the rise of post-Fordism in the 1970s, characterized by increases in flexible workforces and production systems as well as advancements in technology (Jessop, 1993), traditional class delineations became fragmented and popular support of class politics as a catalyst for a more equal and just society declined. Ideas-based movements gained in popularity and the ability to mobilize broader segments of the population were not motivated by calls for socio-economic transformation. The proponents of the organic movement therefore had to shift their goals to accommodate the movement towards 'quality of life' politics that characterized the post-Fordist era (Dalton et al., 1990).

To expand the membership of the movement to the mainstream, the organic movement had to alter its objective from revolutionizing structures and practices to engaging with mainstream society in the 1980s. Engaging with those who were not interested in revolutionizing the agri-food system, but who were interested in health and its links to food, forced advocates for organic agriculture to promote organic food's health claims and its association with a more nutritious diet. The opportunity for the organic movement to gain political momentum largely emerged in response to rising consumer demands for what was perceived of as safer and healthier food across Canada and the US. Mainstream consumers of organic food believed that it was nutritionally superior and healthier than conventionally produced food. But in contrast to the principles emphasized in the process-based definition, the mainstream culture of the 1980s emphasized individual choice and valued the convenience that the industrialized agri-food system offered. One of the benefits of the conventional agri-food system for consumers in North America was its ability to provide a wide variety of fresh foods all year round (Friedmann and McMichael, 1989). Many mainstream consumers who were concerned with food safety, willing and financially able to purchase organic food, were not prepared to compromise their desire for 'food on demand' (Redclift, 1997:336). Mainstream consumers demanding organic foods wanted organic versions of conventional food products beyond fruits, vegetables and grains. They demanded what they perceived of as 'healthier' and more 'nutritious' versions of the processed foods they already purchased. Thus, the 'new culture of consumption' for organic and other types of 'ethical' foods, and the rise in highly individualized food preferences forced the organic movement to adapt to the new political realities of the 'culture of convenience' (Murdoch and Miele, 1999:473): '[The organic sector of the 1980s,] once synonymous with hippies, health faddists and environmentalists, is now sought out by an ever-increasing proportion of mainstream America, people who care about freshness and taste as well as cleanliness and safety' (Burros, 1987).

For the organic movement to gain the acceptance of the general public, its goals and some of its values had to be compromised. Instead of demanding members conform to the more radical objectives of the early movement that rejected consumer interaction with the conventional agri-food sector and norms of industrialized agriculture, the diversity and subjectivity of the membership was opened up to those who politically supported organic agriculture through purchasing organic food, but did not practice organic agriculture or subscribe to a process-based definition of organic. Because of the shift in objectives of the organic movement, organic food's public image in the mainstream media dramatically improved. Thus, what was necessary for the organic movement to keep its momentum and gain membership from the mainstream is what della Porta and Tarrow call 'reframing the issue' (2005:3). Reframing the goal or issue of a social movement goes beyond merely shifting focus to reworking how the idea itself is framed and promoted to the public. In the case of the organic movement, as the mainstream market became the mobilizing structure, the goals of the movement itself placed more emphasis on issues of food quality, safety and environmentalism. Consuming organic food (rather than producing it or eating locally grown products) was promoted as the primary means of supporting the organic movement and the best way of changing the industrialized agri-food system. Essentially, the organic movement reframed the issue to put the responsibility for change in the hands of the consumer while limiting the overall goal of expanding organic agriculture, without radically questioning the form of economic organization associated with industrialized agriculture.

Numerous studies on social movements have explored the issue of cooptation as movements mature and attempt to expand (for example, Lacy, 1982; McAdam, 1983; Coy and Hedeen, 2005). Cooptation generally occurs when challengers of the mainstream alter their claims so that they can be pursued more widely (Meyer and Tarrow, 1998:21). To expand, social movements are sometimes forced to alter their claims to gain attention from the state to achieve some degree of political change, what Cohn has characterized as a 'liberal strategy' (Cohn, 2003:359). The issue of institutionalization in the environmental movement is explored by Forbes and Jermier (2002), which shows that the intention behind institutionalizing (stimulating policy change by engaging with the 'establishment') rarely results in meaningful policy change. Opposed to the more critical stance of the earlier organic movement, during the 1980s the movement shifted its issues and goals to accommodate mainstream tastes to work towards affecting political institutions through more organized action, which has been shown in other cases to result in cooptation and demobilization (Piven and Cloward, 1977). The shift in strategies creates what Belasco calls 'the crossover dilemma', a pitfall encountered by social movements seeking to maintain their momentum by pitching their goals to the mainstream. When a movement seeks to accommodate the interests of the mainstream, the supporters of the original goals are often alienated from the movement because of the compromise of those goals (Belasco, 1989:93). The modification of goals and objectives of the organic movement helped it to bifurcate into the radicals, who still identified with the earlier movement's goals and the process-based definition of organic, and the moderates, who embraced institutionalized politics in an effort to gain some sort of meaningful access to political institutions. The moderate contingent includes actors like businesses that wanted market expansion for organic food and therefore needed government recognition to partake in more expansive trade networks. By emphasizing the qualities of organic food irrespective of the principles and values of organic agriculture as the way to gain broader support through market expansion, the incorporation of organic agriculture into regulatory frameworks was viewed as a way to legitimize organic practices.

Despite the shift in focus of the organic movement to gain broader appeal, supporters and practitioners of the process-based definition continued to invest in 'alternative food networks' to avoid participating in the conventional agri-food market (Whatmore, 2002). Community Supported Agriculture (CSA) networks became an important part of the organizational structure of the organic movement in the late 1980s in North America. Originating in Europe, they were created to establish socio-economic links between local agricultural producers and local consumers supported by the sustainable agriculture movement to advocate change in the agri-food system (Powell, 1995:122). CSAs require members to pay an annual fee to help farmers and growers financially sustain themselves in a competitive market economy. In return, members have access to the farmers' and growers' products once they are harvested. There are no guarantees as to the yield or quality of the products but these networks are primarily focused on providing financial stability to farmers and rural communities while fostering local food chains. These efforts intend to help reduce their exposure to the financial pitfalls of participating in conventional agricultural markets plagued with fluctuating prices based on global supply and demand trends.

CSAs premise their market interactions on direct distribution schemes and local food links. As Powell notes, 'the idea behind all of them is to provide growers with a guaranteed market for their produce, and to give consumers access to a food at a reasonable price. Usually growers and consumers ... live within a short distance of each other, and there may be social links as well' (Powell, 1995:122). The preference by some members of the organic movement for unprocessed organic foods (fruits, vegetables, grains) at the local level has ensured that some element of the movement remains committed to the process-based definition of organic (Goodman, 2000). However, the rising mainstream consumer demand for organic foods in the 1980s radically changed the composition of the movement's membership and allowed for conventional business practices to be included under the banner of the organic movement.

CONCLUSION

This chapter shows the important link between the emergence of the organic movement and other countercultural social movements influenced by the rise of post-materialist values among the postwar generation. Despite the rejection of organic agriculture by the mainstream in the early formation of the movement, as lifestyle choices became an important issue to many, organic food gained appeal among those concerned with food safety and health issues. The political and economic contexts emerging in the 1980s had a transformative effect on the organic movement. In many ways, the organic movement's promotion of organic food in the mainstream market became an important factor in the shift from focusing on materialist goals associated with 'old' social movements. As the mainstream became an important source of support for the movement through organic food purchases, the demand for radical change in individuals' lifestyles as a prerequisite for membership became an unrealistic goal. The mainstream population's interests in consuming organic products for their perceived health and nutritional benefits had grown accustomed to convenient, fresh food and demanded organic versions that appealed to highly individualized preferences. Thus, the product-based definition of organic gained important ideological ground throughout the 1980s in the effort to expand the practices of organic agriculture.

7. New actors, new directions: the contemporary organic movement as an advocacy network

INTRODUCTION

To some degree, two constituencies have always existed in the organic movement. One, consisting mainly of organic producers, has put far more emphasis on the substantive socio-economic and ecological goals associated with the process-based definition of organic, while the other, primarily including consumer movements, focuses on the environmental and health benefits of food produced without synthetic inputs. Yet as the 1980s progressed and the market for organic products expanded, another constituency emerged that deviated from the objectives of independence and self-sufficiency. The establishment of professional organizations bridged the gap between the radicalism of the counterculture of the 1960s from which the organic movement emerged and the new political and economic realities of the post-industrial, consumer culture of the 1980s. The corporate constituency promoting the expansion of organic agriculture while encouraging the institutionalization of organic production processes into the mainstream agri-food system reshaped the organic movement.

This chapter traces the evolution of the organic movement from the mid-1980s until the 2000s. The preoccupation of process-based advocates with banning off-farm inputs such as chemical fertilizers and pesticides made vulnerable the linkage between practice and the socio-economic priniciples associated with the process-based definition. This facilitated a shift in focus from emphasizing the importance of economic and social processes to the value of the end material product, to emphasizing the perceived consumer benefits associated with eating organic foods. This period of the movement's trajectory is distinguished from its previous incarnation by the entry of corporate actors from the conventional agri-food sector. These actors began to advocate for the expansion of organic agriculture by producing and promoting organic foods demanded by the mainstream consumer, while de-emphasizing the association between organic foods and some of the practices promoted in the process-based definition. This time is also characterized by a shift in focus from the social and environmental goals more rigourously promoted at earlier points in the movement's history. Despite the continuing presence of a contingent of advocates committed to practicing the process-based definition of organic, the entry of new actors into the organic movement transformed it from what could be described as a 'social movement' into a transnational advocacy network (TAN). Building upon the policy process model explored in Chapter 6, the TAN model helps explain how issue-based movements extend beyond national political boundaries as new actors not typically associated with social change advocate for a cause.

THE RISE OF PROFESSIONAL ORGANIZATIONS IN THE ORGANIC MOVEMENT

Studies in social movements observe that as movements mature and become established, there is a degree to which the organization of the movement becomes institutionalized in mainstream society. Meyer and Tarrow argue that 'classic social movement modes of action may be becoming a part of the conventional repertoire of participation' (1998:4). In many ways, social protest has been institutionalized in post-industrial societies, such that social mobilization has become a part of modern life, what Meyer and Tarrow term the 'movement society'. Better access to information and media has made today's society much more aware of social, political and environmental issues, including food safety. Reed's (2010) study of the global organic movement shows that the contemporary organization surrounding organic agriculture continues to exhibit the characteristics of a social movement. He claims that the organic movement began as an international form of resistance to conventional and industrial agriculture with the goal of promoting and expanding organic agricultural principles and practices, and ultimately changing the way humans relate to nature. Despite significant changes to its context and membership, it continues to be, Reed argues, a 'cultural social movement' seeking new ways of structuring the conversations surrounding food issues (2010:10). He further argues that the contemporary organic movement 'does not aim to win over government or enshrine itself into law' (Reed: 2010:3).[1] Yet, the organic movement's status as a textbook

---

[1]   Reed's study focuses primarily on the United Kingdom (UK), and to a lesser extent the US.

social movement is called into question as governments and corporations now count themselves among advocates and supporters of organic practices and principles championed by the movement that traditionally rejected their involvement.

As the number of professional business associations supporting organic agriculture in the mid-1980s grew, especially in the Canadian and American contexts, the organic movement's objectives moved beyond persuading mainstream consumers to engage in alternative market relations, to inserting organic agriculture into the mainstream market without demanding broader structural changes to how food is produced, processed and distributed within the conventional market. This phase of the movement no longer presented an alternative to the status quo and instead moved closer to promoting the expansion of the organic market into the conventional agri-food system and its conformance with the principles of global trade. This shift in interests and objectives of the broader movement problematizes the ability to put principles associated with the process-based definition of organic into practice. The impetus behind the expansion of organic practices into conventional structures hinges on the institutionalization of organic into conventional decision-making structures so that organic foods are part of covential markets transactions, yet continue to be differentiated from conventional foods by their credence attributes. Though the movement's promotion of organic food as more nutritious and safer than conventional food has always been a major component of the movement's message, the association of actors who were once the target of protest with the movement's objectives and goals signals a significant change in the structure of the movement and a departure from the movement's earlier political objectives. Thus, the advocacy network model more aptly represents the present form of social organization surrounding organic agriculture in the North American context.

What are the key differences between a social movement and an advocacy network? One important distinction is how policy is included in the analysis of activities in the social sphere. The concept of an advocacy network is used to explain how policy change occurs, and specifically how actors' beliefs and common understandings of the problem influence the policy process (Sabatier and Jenkins-Smith, 1993). But the concept has also been used to explain how groups connected by similar interests are able to meet their collective goals across diverse political landscapes (Keck and Sikkink, 1998). According to Sikkink, social movement studies often place too much emphasis on the structure of a movement in determining its objectives and mobilizing structures, whereas the advocacy network approach argues that the agency of members is highly influential in determining objectives and organizing campaigns (Sikkink, 2005:151). What truly differentiates an advocacy network from a social movement is the effort made by members of the network to influence their immediate political situation by involving themselves in political institutions (Keck and Sikkink, 1998:4). From this agent-based perspective, actors in an advocacy network have the ability to exert pressure on organizational structures to stimulate change in objectives or the way resources are distributed within their organizational network through interactions with political institutions outside of the network. This does not mean that social movements are fundamentally different from advocacy networks, but Keck and Sikkink's model gives more authoritative weight to the behaviors of actors within the structure, compared to the social movement model's emphasis on the way structures impose limitations on actors' behavior.

In some cases, advocacy networks and social movements share characteristics including ideas, objectives, membership characteristics and ways of distributing resources, and may even compete with each other for legitimacy (Keck and Sikkink, 2000:217). Members in networks are often diverse and can include not only NGOs found in more traditional social movements, but a range of other members including businesses and governmental organizations (O'Brien et al., 2000:110). However, despite the more agent-centred approach of the advocacy network, power differentials exist and can be exacerbated to a greater degree than in a traditional social movement. Often the interests of stronger actors in the network tend to drown out those of the actors with less access to financial or political resources (Dalton, 1990:27).

Applying the classification of advocacy network to the organic movement is not new to studies in organic agriculture, and it appears to be particularly useful when characterizing the organic movement in the twenty-first century. In Bostrom and Klintman's 2006 comparative study of organic regulatory frameworks in Sweden and the US, they identify the existence of two constituencies with varying degrees of interest in the expansion of principled practices associated with organic agriculture. Bostrom and Klintman's definition of an organic advocacy network, which as they claim defines the current form of organization of those with an interest in the expansion in all forms of organic agriculture, is understood to include 'all organizations and individuals that actively support and are engaged in organic production in any way' (Bostrom and Klintman, 2006:167). This is a very broad understanding of what constitutes the organic advocacy network, but it serves to recognize the significant changes in the membership of the movement beginning in the 1980s. The type of social organization used to describe how issues

related to organic agriculture are transmitted to society must acknow-
ledge that materialist concerns do not play as prominent a role among the
current group of actors campaigning for organic agriculture's expansion
as they did in the past.

The membership in the broader organic advocacy network does not
necessarily subscribe to the social, economic or environmental principles
associated with the process-based definition of organic, or those historically associated with practices in organic agriculture (Sumner, 2005). As
observed in other advocacy networks, once common principles are
established, they bring members together who at earlier points interpreted
the principles differently (Keck and Sikkink, 1998:36). Because of the
existence of two constituencies in the organic advocacy network – one
which promotes the process-based definition of organic and the other
which promotes the product-based definition – the commitment of
members to the social and economic principles of organic agriculture
varies considerably. What links all members is the shared dedication to
organic agriculture's more instrumentalist principle of no synthetic inputs
that impact the material qualities of organic foods.

As the organic movement evolved in Canada and the US in the latter
part of the twentieth century, the number and diversity of the actors who
associated themselves with organic principles multiplied. Instead of
'Back to Landers' and 'middle-class hippies', the developing market
potential for organic food drew in many interested parties seeking to gain
from the growing politicization of food issues and organic agriculture's
niche market status. Although the organic movement has always valued
the institution of the market as an effective mechanism to spread organic
agriculture and transmit organic values to the public through consumption (increased demand stimulating supply), some of the members of the
organic movement who joined in the mid-1980s prioritized the expansion
of the market for organic products over the more substantive goals
associated with the process-based definition of organic. The early organic
movement's dependence on trust-based relationships to assure fair deal-
ings in market interactions made it easy for outside actors to enter the
organic market as the acceptability of organic grew in the mainstream
and consumer's willingness to pay for organic food was realized. The
consequence of focusing advocacy efforts on restricting off-farm inputs,
the only principled organic practice that influences the material qualities
of organic food, was that less attention was given to putting into practice
the other principles of social and ecological sustainability. In 1976, the
book *Radical Agriculture* assembled a range of voices critical of the
developments in agriculture in North America. In one chapter, radical

ecologist Richard Merrill reasoned that conventional agricultural practices would be excluded from the expanding organic movement because
governments continued to see the purpose of agriculture as producing
high yields. This type of fossil fuel intensive production can only operate
'efficiently in an industrial mileu' (Merrill, 1976:xix), making conventional practices unlikely to be taken up by organic advocates. Other
authors argued against the notion that conventional agricultural practices
could become part of the organic movement:

> to be a technical connoisseur of an 'organic' approach to agriculture is no
> better than to be a mere practitioner of a chemical approach. We do not
> become 'organic farmers' merely by culling the latest magazines and manuals
> in this area, any more than we become healthy by consuming 'organic' foods
> acquired from the newest suburban supermarket. (Goldstein, 1976:10)

Yet, at this time active business associations were advocating for organic
agriculture that prioritized the expansion of markets for their products by
utilizing practices found in the conventional agri-food system. Businesses
interested in expanding organic production were eager to engage with the
state as conventional agri-food companies did, to set regulations that
would further facilitate market growth and trade in organic products. The
new professionalized, business-oriented constituents, referred to as 'the
pragmatists', drove the institutionalization of the organic movement and
helped to transform it into an advocacy network that promoted market
expansion through the promotion of organic foods sales to mainstream
consumers (Egri, 1994:151). However, this shift in focus did not happen
all at once and cannot be attributed to one single event or actor within the
movement. Instead, there were three major turning points in the shift in
focus of the movement from the promoting of the process-based definition to the product-based definition of organic. First, the move to create
enforceable certification structures that extended beyond local organic
agriculture chapters signaled a significant change in the way organic
value chains were monitored and managed. Second, the growing public
concern with food safety drew broader attention to organic food as more
nutritious and safer than conventional food, helping to brand organic food
as part of a healthy lifestyle. The third shift was towards engagement
with the state to work together to protect the integrity of the organic label
and market share through regulatory enforcement of certification standards. These shifts saw new groups of actors emerge as advocates for
organic food. Though the interest of consumers and national governments
in organic food were two important factors in the changes observed in the
organic movement, the emergence of multiple sets of objectives within

the movement were pivotal in the transformation of the movement into an advocacy network.

As argued by Cox (1999:10–11) and O'Brien et al. (2000) in their discussions of globalization and civil society, there are three general groupings within social movements that emerged in response to changing socio-economic structures in the global economy in the 1990s: conformers, reformers and radicals. Conformers are defined as accepting the status quo in regard to the global trade system, while reformers want some degree of change to the way the global system of trade functions. The radicals completely reject the system and characteristically want the global trade regime dismantled or replaced with something new. The radical contingent can be understood to be associated with the broader anti-globalization movement. Organizations within the broader organic advocacy network tend to fall into one of the first two of these three categories in terms of their orientation towards the global agri-food system. Conformers and reformers tend to be more professionally organized than the radicals. As Guthman (2004:111) points out, as many of the loosely knit networks made up of organic producers' associations matured, they transformed into more professional trade organizations and certifiers as the market for organic food expanded, and state legislation became necessary to faciliate its further growth.

One of the most influential and powerful associations that can be considered a conforming organization in the organic advocacy network is the Organic Trade Association (OTA). First established in 1985 under the name Organic Foods Production Association of North America, it changed its name in 1994 to account for the portion of its membership involved in the value chain beyond the production stage. It includes some of the largest organic businesses in both Canada and the US including SunOpta, Whole Foods Market, the Hain Celestial Group and United Natural Foods (OTA, 2014). The OTA is a transnational association linking businesses on both sides of the border through their mutual business interests. Its membership now draws from all sectors including producers and retailers. It represents 'the industry's interests to policy makers, the media and the public' (OTA, 2006).

The OTA's mandate is to promote organic business interests, which includes protecting the growth of organic trade to 'benefit the environment, farmers, the public and the economy'. The OTA receives funding from its membership which currently includes some of the biggest investors in the organic food industry, such as Cascadian Farms and Campbell's Soup (OTA, 2013). Its activism was highly influential in getting organic regulations on the agenda of policy makers in both Canada and the US, and helped politicize the importance of labeling and certification schemes of organic products. It was an important source of perspective and information for both the American and Canadian governments as they devised their national organic standards and entered into bilateral equivalency arrangements. The OTA has helped the organic food industry gain legitimacy in the mainstream through its lobbying of governments for organic regulations and labeling in addition to its willingness to embrace conventional business practices and agri-food corporations like the Hain Celestial Group.

Another organization active in the organic advocacy network that can be considered a conformer is the Organic Federation of Canada/ Fédération Biologique du Canada (OFC/FBC). It was created in 2007 and has an advisory role with the Canadian Food Inspection Agency (CFIA) since the ratification of the Canadian national standard in 2009. Membership includes the Canadian OTA and provincial/territorial level associations representing organic producers, processors, retailers and consumers. It meets with industry and the CFIA to maintain open dialogue between regulators, industry representatives and other members of civil society. Its key goals include maintaining open communications with the Government of Canada to assure that national organic regulations remain relevant and up to date. It lobbies on behalf of those involved in the organic industry who wish to see the expansion of Canadian trade in organic products. It also serves as a clearing house to provide information regarding certification and regulations to interested parties. Its key goals include 'reconcil[ing] success with commitment to the principles of organic agriculture', but it is unclear as to what the OFC/FBC includes in the set of principles it references. Though its mandate is to expand and improve market conditions for Canadian-produced organic products, it does not include reforming economic relations in organic agri-food value chains to strengthen more localized consumption of Canadian-certified organic products (OFC/FBC, 2012). Because the OFC/FBC represents such a diverse group of actors, it may appear that divisions exist between members. It is true that not all members share similar approaches to achieving broader goals, but to gain attention from the government, a united front with consistent goals and principles is essential to participation in the formal decision-making process.

An example of a professional association that can be considered a reformer is the OCA. Originally named the Pure Food Campaign in 1992 by its founder Jeremy Rifkin, its name was changed to OCA in 1998 to reflect the focus on organic food and agriculture standards. Its launch in 1998 was spurred on by the United States Department of Agriculture's (USDA) first draft of organic standards that did not exclude irradiation or GMOs from the certified organic label (which have since been excluded;

see Chapter 4). The OCA's president, Ronnie Cummins, took issue with the National Organic Standards as put forth by the USDA (and supported by the OTA) in the early 1990s. It has over 850 000 members including 3000 food cooperatives in both Canada and the US. Its goal is to protect consumers from fraudulent claims made by organic food labels while promoting 'health justice' and sustainability, as well as a number of other related issues such as the fair trade and anti-GMO campaigns. The OCA does not believe that national organic standards go far enough to truly represent the principles of the process-based definition of organic. It sees the current sets of standards as catering to large-scale production in the organic food sector.

The OCA also performs a 'whistle-blower' function by drawing the public's attention to publicized claims of fraudulent organic labels and to companies that do not adhere to the process-based definition of organic. Far more critical of the exemption of more substantive goals from national standards than conformers in the network, the OCA takes issue with the corporate consolidation of the organic food industry and often publicly criticizes agri-food corporations like Horizon Dairy, which almost consolidated the entire organic dairy sector in the US in the 1990s (as discussed in Chapter 3). The OCA proposed the 'Safeguard Organic Standards' policy that would block larger corporations from insinuating in product advertisements that their organic products adhere to the process-based principles when only the minimal certification require-ments are met (OCA, 2006b). The organization encourages consumers to boycott certain companies for their business practices that do not coincide with organic principles, and also pressures national governments and companies to preserve organic standards and lobbies to have them strengthened (OCA, 2007). The OCA, like other consumer associations, believes that the global trade regime privileges the interests of corpor-ations over the interests of consumers (Williams, 2005:42). But while the OCA rejects excessive corporate consolidation of organic agriculture it firmly believes that market mechanisms and the rise of the 'ethical' consumer are the best means of achieving social change. Like other prominent actors in the organic advocacy network, it is mainly concerned with consumer issues in the mainstream market and urges consumers to treat their purchasing habits as a form of political action. As the slogan on its website reads, 'put your money where your mouth is' (OCA, 2007).

While not a transnational organization, Canadian Organic Growers (COG) is a professional organization that can be categorized as a reformer. COG is a member of the International Federation of Organic Agriculture Movements (IFOAM) and is made up of farmers, gardeners, retailers, consumers, policy-makers and educators. It has local chapters across Canada and publishes a quarterly magazine called *The Canadian Organic Grower*. Like the OCA, COG supports the general goals of sustainability and social justice, and has taken issue with the conventionalization of organic production processes. Unlike the OTA or the OCA, COG's mandate includes bioregional organic agri-food systems, emphasizing the importance of localized organic value chains to the sustainability of the agri-food system in Canada (COG, 2014). In *The Vancouver Sun*'s series on organic agriculture, COG's executive director Laura Telford lamented the shift in focus of the organic movement, stating 'my biggest problem is … they're not really adopting organics in the way our organic pioneers had imagined' (Weeks, 2006:D4). While social justice and sustainability are supported in principle, like other professional organizations included in the organic advocacy network, COG does not have a mandate to pressure its policy-making members to address the importance of social issues in organic regulations and standards. In 2012, COG published a book called *Scaling Up* that included interview material from organic farmers across Canada. The book promotes a better understanding of the successes and challenges of farming organically in Canada. It raises the 'taboo' questions of organic agriculture related to efficiency and scale. The book points out that developing ways to 'build and manage pests and diseases while adhering to organic principles is only part of the journey towards sustainability. True sustainability includes environmental, social and economic elements' (COG, 2012:1). This statement largely captures COG's approach to organic agriculture and what the organization sees as the real value of expanding organic production. Though COG does not have an official 'social standard', the organization is committed to 'leading local and national communities towards sustainable organic stewardship of land, food and fibre while respecting nature, upholding social justice and protecting natural resources' (COG, 2014).

Professional associations like OTA, OFC, OCA and COG have been actively involved in policy agenda setting but do not call for radical structural changes of the globalized agri-food system as earlier members in the organic movement once did. The radicalism associated with organic agriculture in the 1960s is not a large part of the professionalized organic advocacy network as it exists today, and does not exist in organizational mandates in a distinct way. What Rodale, Howard and their contemporaries championed was a radical overhaul of the way food was produced and distributed. There are of course, notable movements and networks seeking radical material reorganization of the globalized agri-food system, such as La Via Campensias (International Peasant

Movement), food sovereignty movements and the Food Chain Workers Alliance, who seek social justice for farmers and workers in the agri-food system. However, these activists do not exclusively promote organic agriculture or organic food as the more desirable alternative to the conventional system.

THE PROFESSIONALIZATION OF THE ORGANIC MOVEMENT

The professionalization of organizations within social movements tends to shift the focus of the broader movement, often eroding the original goals (Tilly, 2004). Similar to what Tilly argues in his examination of social movements, as the organic advocacy network has included more professionalized associations, it has shifted its focus away from local and regional interests that were once a vital part of its objectives. This is what Meyer and Tarrow refer to as 'the paradox of professionalization'. Professionalization, as a type of institutionalization, is a formalization of interests and membership under the guise of a consistent mandate. It allows for networks in social movements to have access to decision-making institutions to further their mandate in hopes of achieving a degree of meaningful social change in the form of new policy. Professionalization, although necessary to gain recognition and legitimacy from decision-makers, may eventually undermine sustained efforts at mobilization because the goals of the organizations shift from collective action at the grassroots level to influencing policy (Meyer and Tarrow, 1998:15).

The inclusion of professionalized organic associations, including business and consumer associations, has helped to change the characteristics of the organic advocacy network so that it reflects a wider range of participants in the organic sector that includes corporate actors traditionally the target of social protest. As observed in the evolution of the environmental movement, the 'establishment complex' is often triggered when business interests, considered by the non-governmental organization (NGO) community as the establishment, start promoting the values once associated with social movements (Hodess, 2001:140; Cox, 1999:11). The fluid and less restrictive criteria for membership in the organic advocacy network weakened the policing of further claims of membership, largely reducing it to adherence to the product-based approach to organic production methods which, as discussed earlier, said very little about preferred types of market behaviors and economic activities. As Bostrom and Klintman argue, 'as a result of an institutionalization of movement practices, the advocacy network can continue to provide stimuli for a reformed conventional agriculture, while the movement has lost the capacity to fundamentally challenge it' (2006:171). Although the more substantive issues associated with the process-based definition remain the concern of some in the organic advocacy network, they are not major issues in current campaigns. Associations such as the OCA and COG agree with fair labor standards in principle, but social issues have not been anywhere as important as market expansion in official policy recommendations.

Food Safety

Professional members in the organic advocacy network were influential in altering the objectives of the network as presented to the public, creating what Keck and Sikkink call 'information politics'. Advocacy networks 'relay ideas in order to alter information and value context' to mobilize the public (Keck and Sikkink, 1998:2,16). Mobilizing information creates new issues and categories that grab people's attention by changing their perception of the issue. Some members of the organic advocacy network drew attention to the dangers of the industrialized agri-food system to convince people that organic foods were safer because they did not participate in a system that increases the risk of health and safety hazards. Food safety gained more public attention in the US and Canada after a number of scares and recalls emerged in the late 1980s and throughout the 1990s (for example, the Alar food scare, E.coli and salmonella contaminations) (McGrath, 1991; Schlosser, 2002; Nestle, 2004). In the 1989 edition of the produce industry's paper *The Packer*, the headline reads 'Product Safety: Organics – Hot Demand, Short Supply', which discusses how the growth of the organic food sector was largely in response to fears about foods produced in the conventional system (Friedland, 1994:225). The organic advocacy network used the media attention to food safety issues to promote its objective of expanding organic agriculture. It did this by promoting the idea that organic production methods lessen food safety concerns because they do not use industrial chemicals in their production processes and should therefore be the preferred way of producing food (Soil Association, 2001:5). Concerns over food safety even prompted the recruitment of celebrities to help publicize the objectives of the network. In a 1989 *Organic Gardening* (*OG*) magazine feature, actor Meryl Streep was interviewed about her personal campaign against high levels of agricultural chemical residues in children's food (*OG*, 1989). Conducting public opinion polls has also been an important part of mobilizing the public through the dissemination of information. Citing the 'Fresh Trends' consumer poll in *The*

Packer, a 1989 issue of *OG* reports on the growing public concern about food safety issues. More than 80 percent of respondents had concerns about the safety of their food, which led 18 percent to change their food purchasing habits (*OG*, 1989:43).

Linking organic food with food safety issues was an effective way of linking issues and promoting organic foods to mainstream consumers. In a survey conducted by *OG* in 1989, 48 percent of those surveyed 'regularly' ate organic fruits and vegetables, and over half of them chose organic for its health benefits (*OG*, 1989:42–6). Thus, the promotion of organic agriculture was reframed as a human health issue linking chemical residues on food with food safety, which linked organic food with other consumer movements. Linking the organic and food safety issues qualify as what Bennett labels 'loose activist networks' which overlap multiple issues and goals, and possess flexible member identities (Bennett, 2005:213). A more recent association is the linkage of the organic advocacy network with the anti-GMO movement.

The Anti-GMO Movement

As genetically modified organisms (GMOs) in the agri-food system received growing negative publicity in the 1990s, the organic advocacy network began to link its objectives to the anti-GMO movement (Reed, 2010:110). Biotechnology, as defined by the Cartagena Protocol on Biosafety (CPB), is 'any technological application that uses biological systems, living organisms, or derivatives thereof, to make or modify products or processes for specific use'. Modern biotechnology is defined as 'the application of in vitro nucleic acid techniques, including recombinant deoxyribonucleic acid (DNA) and direct injection of nucleic acid into cells or organelles, or fusion of cells beyond the taxonomic family' (CPB, 2012). This technology has been used to create herbicide resistance in plants (notably canola, soybean and corn), as well as plants with pest resistance qualities such as Bt corn. These crops are commercialized and used in the agri-food system as well as for industrial uses; for example, canola oil is used as an industrial lubricant. The US and Canada are the leading producers of genetically modified (GM) crops along with Brazil, Argentina, China and South Africa. By 1998, it was reported that 30 percent of all soybeans grown in the US and over one-quarter of all maize was genetically modified. This number jumped to 45 percent of all maize in 2003 (Singer and Mason, 2006:208). In 2012, 85 percent of all soybeans cultivated in the US were genetically modified. The EU imports a large portion of GM soybean grown in the US for livestock feed (GMO Compass, 2012). Neither Canada nor the US requires producers to label food products containing GMOs as they have been scientifically assessed by various government agencies responsible for food safety (for example, the CFIA, or the US Food and Drug Administration (FDA)) to pose no risks to human, plant and animal health.

Commercialized plants that are genetically modified are most often developed by large transnational agribusinesses such as Monsanto, Bayer and Pioneer Hi-Bred, though some public institutions are involved in the research and development of GM varieties. The use of crops that utilize biotechnology is controlled by patents/user fees, which are passed along to producers. Like inputs derived from fossil fuels, biotechnology is viewed by those who oppose the use of the technology in the food system as a product of industrialized agriculture. GMOs are also perceived as technologies that are potentially dangerous to human and environmental health, and carry unknown future consequences. The use of GMOs is therefore banned in 'certified' organic agriculture. Scholars such as Kinchy (2012) argue that it is not solely the 'science-based' safety record of GM crops that is the issue for organic practitioners and other anti-GM supporters. Rather, it is the 'scientization' of governance that has limited the debate over biotechnology to matters of risk and safety while silencing the ethical, environmental and social considerations pertaining to the proliferation of GM crops in the agri-food system.

The anti-GMO movement emerged in reaction to the growing control of agribusiness in the genetic manipulation of certain crop traits and the perception that GM foods were being clandestinely incorporated into the agri-food system. There were claims made by some scientists that GM food products carried a significant level of unknown risks regarding future harm to humans and the environment. The fear surrounding 'frankenfoods' mobilized people against GMOs in the agri-food system that were not previously affiliated with the organic food or environmental movements (Reynolds, 1999). The growth of the anti-GMO movement helped direct those concerned with the biosafety of the agri-food system towards the principles and mandate of the organic advocacy network.

The movement was able to frame the biotech issue to mobilize public support against GMOs with a number of highly publicized legal cases. One such case occurred when Monsanto launched a lawsuit against Saskatchewan farmer Percy Schmeiser for violating intellectual property law when traces of 'Roundup Ready' canola was found in his fields in 1997. Monsanto claimed that Schmeiser planted Monsanto's 'Roundup Ready' canola seeds in 1998 without paying the patent fees (since 'Roundup Ready' is genetically altered and its use is patent-protected). Schmeiser himself claimed he did not plant the seeds and that they were blown over from an adjacent farmer's field (Schmeiser, 2007). Schmeiser claimed that Monsanto's seeds had polluted the strain of canola he had been breeding and developing for 50 years. In 2004, Monsanto won its case against Schmeiser at the Supreme Court of Canada, although Schmeiser did not have to pay damages to Monsanto (www.percy schmeiser.com). Another case, which drew the public's attention to GMOs, occurred in 2000 when it was reported that a GM-strain of corn (Aventis' Starlink brand) not meant for human consumption was discovered in the US food system (Borenstein, 2005). The Bt corn contained the *Bacillus thuringeinsis* toxin, a type of soil bacteria inserted into the corn's genetic material that helps the plant resist pests. The corn was not approved for direct human consumption because it could possibly trigger reactions in consumers with allergies to the Bt protein (Nestle, 2003: 2–3). As a result, there was a massive food recall of taco shells and other corn products that possibly contained the Starlink corn. Both cases publicized a number of issues the anti-GMO movement addresses: the power agribusiness has over the activities of farmers and the lack of control corporations have over the unintended and potentially harmful commingling of GM products with non-GM products in the agri-food system (Kinchy, 2012). The current anxieties about health and the environment linked to the anti-GMO movement are intertwined with issues surrouding power and inequality in the agri-food system. Advocates argue that these issues cannot be reduced to scientifically determined probabilities of risk.

The anti-GMO movement is a diverse group of members including the National Farmers Union, environmental groups like Greenpeace, Canadian Biotechnology Action Network (CBAN), FoodDemocracyNow!, Friends of the Earth as well as concerned individual scientists and citizens (Council of Canadians, 2005). Some anti-GMO groups want GM foods to carry labels, while the more radical constituency calls for a ban of biotechnology in the agri-food system all together (Nielsen and Anderson, 2000). Mobilizing against GMOs has been politically effective in Europe where the EU has restricted the use and circulation of GMOs in the agri-food system under its Directive 2001/18/EC legislation (only GM maize has gained approval for cultivation in the EU), while some food retailers in Europe have gone completely 'GMO-free' (Bullock et al., 2000). It should however be noted that some GM crops, namely soybeans, are imported into the EU (primarily from the US and Brazil) and used for livestock feed (the resulting animal products do not carry GM labels).

The impact of the anti-GMO movement in Canada and the US has not resulted in an outright ban of GM technology in the agri-food system, although public pressure continues to mount in an effort to legislate mandatory labels on food products containing GM ingredients (Andree, 2011). Ironically, some members of the organic food community do not support labels on GM ingredients in foods, as this has the potential to undermine the value of 'certified' organic products as the only 'GM-free' foods on the market. The involvement of particular corporate actors in the labeling GM debate demonstrates the further diversification of the organic advocacy network. In 2012, several companies producing certified organic foods opposed the attempt to pass legislation that would require the labeling of GM ingredients in California. The list of companies includes Horizon Organic Dairy, Cascadian Farms, Muir Glen and Santa Cruz Organic.[2] As of 2014, there are no mandatory labels on GM foods in either country and GM ingredients have not been eradicated from the agri-food system, although several 'GE free' zones have been established in British Columbia (Richmond, Rossland, Nelson and Powell River) at the municipal or regional levels (CBAN, 2012). Clearly, the membership in the organic advocacy network does not have a shared coherent position on the labeling of GMOs in the agri-food system. As more corporate actors enter the organic food sector it is likely to become even more diverse in terms of membership and message relating to biotechnology.

The anti-GMO network recommends that concerned consumers should pressure governments to ban or limit the use of GMOs in the agri-food system and to buy certified organic products because organic certification standards forbid the use of GMOs in organic agriculture. Although GMOs were not referenced in the early organic movement (because they did not exist until the 1980s) today the organic advocacy network rejects GMOs entirely, and was influential in banning the use of GMOs in organic agriculture regulations in the EU and US (Baker, 2004). The network has also been a crucial actor in efforts to lobby sub-state and state level governments to label GMOs in Canada and the US. Linking organic and anti-GMO campaigns through what della Porta and Tarrow label 'transnational collective action' (2005:2–3) helped to boost the organic sector as it was viewed as a beneficial strategy to keep GMOs out of one type of food production. The organic advocacy network and the anti-GMO movement are now part of a larger transnational food issues network. The OCA is an example of an organization that has effectively

---

[2] Proposition 37 which appeared on the 2012 California ballot proposed mandatory labels on all food products sold in California that contained GM ingredients (with some notable exemptions such as California wine). This proposition did not pass, but other states (Oregon, Vermont) have attempted to pass similar state-level legislation (Clark, Ryan, Kerr, 2014).

linked these two issues as part of their mandate which includes a 'global moratorium on genetically engineered food and crops' (OCA, 2012).

The organic and anti-GMO movement share two goals – to bring public attention to the negative environmental and human health effects of the practices of conventional agribusiness, and to link the practices in conventional agriculture with food safety issues. Another element that strongly links these two activist networks is the focus on the content of the end material product. As they have formally united these goals in their mandates, both efforts have gained broader public support and attention from governments. As Keck and Sikkink demonstrate, 'advocacy networks have been particularly important in value-laden debates ... where large numbers of differently situated individuals ... have developed similar world views' (1998:9). By highlighting the similarities between networks they are equally able to get their message out to the public and gain broader support.

'Progressive' Food Movements

In addition to the anti-GMO movement, the organic advocacy network is part of a broader food movement which includes a diverse group of activist networks geared towards specific goals and objectives that involve change to the current organization of the agri-food system. The broader movement currently consists of networks primarily focused on challenging the production and distributional aspects of the globalized, industrialized agri-food system. The rise of the 'ethical eating counter trend' helped the progressive food movement to gain momentum as consumers began to associate the purchasing of 'ethical foods' with political action (Bell and Valentine, 1997). The Slow Food Movement, locavorism and the '100 mile diet' for example, contest the 'just-in-time' model of food production associated with globalization and transnationalized value chains (Pollan, 2009; Smith and MacKinnon, 2007; Petrini and Watson (eds), 2001; Kerton and Sinclair, 2010). They encourage consumers to make food choices that reject the industrialization of food production either by purchasing foods grown or processed locally, or to stop purchasing highly processed foods that move profits from food sales downstream from farmers and other producers. These causes promote the link between food quality and healthfulness with less industrial forms of food production.

The food sovereignty network also challenges the structure of the global agri-food system by advocating reform of the current system of trade, and its treatment of food and agriculture (Goodman and Dupuis, 2002:19; Miele and Murdoch, 2002; McMichael, 2004; Desmarais et al.

(eds), 2011; Andree et al. (eds), 2014). It also embodies democratic principles as it advocates for dissaggregated control over land and resources related to agricultural production. It is often associated with the organic advocacy network because of the ongoing linkage between organic agriculture and some of the principles of the process-based definition of organic, in particular localized value chains. Food sovereignty networks include organizations such as Food Secure Canada and the International Planning Committee on Food Sovereignty, that campaign for local and regionalized agri-food systems, dissaggregate and democratic decision-making structures in agricultural sectors and the movement away from the global agri-food system organized around the principle of comparative advantage and corporate models of ownership. Some have taken issue with linking the idea of social change as championed by the food sovereignty network with the expansion of organic agriculture and its involvement in transnational value chains. As Allen and Kovach state in their exploration of the social implications of organic food entering the mainstream market, '[it] requires collective action in the form of a social movement, not the "invisible hand" of the market' (Allen and Kovach, 2000:230).

The organic advocacy network also links its objectives with the fair trade network and dual labeling schemes, as 'certified organic' and 'fair trade' appear on an increasing number of food products because both causes use labels to represent products that have credence attributes. Linking the fair trade and organic labels on products is largely driven by consumers' demands for food products that reject some aspect of industrialized agriculture. The original motivation behind the fair trade network was to address exploitive labor practices in the globalized economy and the highly volatile international market for commodities such as coffee, bananas and cocoa grown primarily in developing countries in Latin and South America (Guthman, 2004:119). Fair trade has a much closer relationship to the process-based definition of organic with its emphasis on social benefits as opposed to the product-based definition that largely inhabits the contemporary organic advocacy network. Both networks have experienced institutionalization to some degree (Gendron et al., 2006; Ransom, 2005; Fridell, 2007) as they continue to engage with the state to gain legitimacy and recognition through government-sanctioned, enforceable standards for their labels. Despite the fact that some members of the organic network have joined forces with the fair trade network, most popularly in the production and labeling schemes for coffee and cocoa through alternative trade organizations, the issue linkage between the organic and the fair trade campaign is not as strong as the issue linkage between the anti-GMO and organic networks (Blowfield, 2001) because fair trade's certification is based on achieving a particular type of social condition, which is an issue of process. However, like the anti-GMO campaign, the organic advocacy network and the fair trade campaign focus on influencing the purchasing habits of consumers to meet their goals through the use of certification and labeling schemes.

The world of progressive food movements has expanded since the 1990s to include many types of activism and movements premised on social and ecological justice that are linked across borders. This chapter is by no means an exhaustive account of the food movement *per se*, but rather highlights the links between movements that influenced and helped to change the organic movement.

CONCLUSION

Organic agriculture is often associated with environmentalism and sustainability, and is understood to have a close ideological connection with other social movements seeking socio-political change to the status quo (Allen and Sachs, 1991). Social movement theory is useful in assessing the shape of the organic movement when its membership was poised to challenge norms associated with industrial agriculture, as other movements like the Slow Food movement and food sovereignty network continue to do. The shift in structure of the organic *social movement* to the organic *advocacy network* has been accompanied by a transformation in objectives that has limited the political traction of organic agriculture and its ability to mount a challenge to industrialized forms of food production and the institutions that manage them. The contemporary organic advocacy network consists of a variety of actors with diverse interests that are often linked with other networks. How organic agriculture should expand, what should be included in this expansion and what aspects of the production process should be included in organic regulations is where members of the broad network tend to deviate. The effectiveness of professional business associations in informing the public about how the expansion of organic agriculture will benefit the environment and their personal well-being has helped to reframe the issues of the network and shifted its focus. Professional associations with financial and political resources dwarfing those of the more radical, less institutionalized constituencies have successfully reframed the issue to highlight their particular interests of market expansion and uniform regulation to further facilitate the corporate approach to organic food production.

Some professional associations such as the OCA claim that 'voting with consumer dollars' is the most useful way of helping to expand organic agriculture on a global scale (OCA, 2007). The transition of organic agriculture from a form of political resistance to a type of food production supplying premium-priced food to a niche market for affluent consumers has earned organic food the nickname 'yuppie chow' by those who question the motivations behind its mass market appeal (Guthman, 2003:45). Those with sufficient disposable incomes wishing to make ethical consumer choices can now do so without radically changing their lifestyles. But ethical consumerism is primarily enjoyed by those who have the disposable income to spend on premium-priced food. The broader organic advocacy network continues to struggle to find sustainable ways of making organic food affordable and accessible to more people while advocating for organic producers to receive income that fairly reflects their efforts. Although the 'new progressive food politics', of which the organic advocacy network is now a part, has the objective of changing patterns in conventional agri-food networks through consumer choices, challenging the organization of the global economy extends well beyond food politics.

8. Conclusions – organic limited

INTRODUCTION

In September 2006, organic food's longstanding reputation as healthier and safer than conventional food was in jeopardy. Over a hundred people across the US became sick from ingesting bagged organic spinach tainted with E.coli bacteria that originated from Natural Selection's organic produce operations in California. A 77-year-old woman and a 23-month-old girl died from ingesting the spinach. After being investigated by the US Food and Drug Administration (FDA), Natural Selection was cleared of responsibility for the contaminated spinach. Instead, it was found that improper handling somewhere between the field and the plate was to blame for the contamination (Canadian Broadcasting Corporation (CBC), 2006b). Just a month later across Canada and the US, organic carrot juice was pulled off supermarket shelves because of possible botulism contamination. Four Americans became sick from drinking the tainted carrot juice. One of the makers of the carrot juice was Earthbound Farms, which owned Natural Selection at the time (CBC, 2006a). In another instance, a recall was ordered for over 3000 peanut products including organic products in 2009 because of salmonella contamination that originated in a peanut processing plant in Texas. The Peanut Corporation of America was at the centre of the outbreak, and was allowed to keep its organic certification despite not having a valid health certificate. Nine people died and 700 became ill from ingesting the tainted peanut products (Severson and Martin, 2009). A search of an FDA-affiliated website for food recalls reveals dozens of organic products pulled off grocery store shelves including peanut butter, chia seed powder and ice cream from the early 2000s to the 2010s (foodsafety.gov, 2015). In 2014, three food recalls were issued by the Canadian Food Inspection Agency (CFIA) concerning organic products: one due to salmonella contamination, the second due to plastic pieces found in the product and the third case was due to undeclared milk ingredients in the product (CFIA, 2014). The episodes of tainted organic food products circulating in the North American agri-food system sparked concern across Canada and the US and called into question the safety of organic food products and whether they are indeed safer than conventional. Though the cases of safe organic products on the grocery store shelves far outweigh the cases of unsafe ones, recalls over food contaminations of any kind cause fear and consumer panic, shaking consumer confidence in products that are advertised as safer and healthier no matter whether or not they are organic.

Despite the arguments made by advocates for a globally integrated, transnationally organized organic food industry, the integration of organic food into the conventional food system has its drawbacks. Some involved in the organic food industry are motivated to produce organic products on an industrial scale because of the premium prices so many are willing to pay for organic products. It has been the goal of this book to explain how the product-based approach to organic production has transformed the politics of organic food and marginalized the approach to food production premised on differentiating itself from industrial practices. It has also tried to show that the organic movement's preoccupation with restricting off-farm inputs allowed the link between principles and practice to be reinterpreted by some who see organic food simply as food that does not contain synthetic chemicals or GMOs. As a result, the core element of social sustainability, which is fundamental to the process-based definition of organic, was vulnerable to interpretation and revision in a way that banning sythetic inputs was not. As the market for organic products expanded, so too did the definition of the relationship between social benefits and organic. This has led to the transformation and fragmentation of the organic movement into a loosely connected network of actors functioning in the organic sector, which may individually hold very different ideas of what responsibility, if any, organic agriculture has to social and economic principles that deviate from those found in conventional practices.

Early practitioners and supporters of organic agriculture imagined that an 'organic future' could be realized, which would replace the industrialized agri-food system with a 'post-industrial ecologically sustainable system of family farmers [who] are again agrarian crafts persons' (Egri, 1994:131). The contemporary influence of the product-based definition of organic would have undoubtedly surprised early organic advocates like Howard, Balfour and Rodale. Few could have imagined, even just 30 years ago, the dramatic changes observed in the organic agri-food sectors in the US and Canada. The processes of corporatization and institutionalization of the organic movement have had far-reaching consequences for re-imaging the global food system. Though the process-based definition was not fully realized as early practitioners envisioned, it did lay important groundwork for advocates of a more socially and ecologically sustainable agri-food system who continue to link organic agriculture with social responsibility.

THE CHANGING POLITICS OF THE ORGANIC MOVEMENT

This book has attempted to show that a signficiant contingent of the organic sector in terms of its structure, principal actors, ideas and institutional frameworks has changed to such a degree that it no longer resembles early understandings of organic agriculture as a challenge to the practices and principles of industrialized forms of food production. By examining the two distinct definitions of organic, the economic dimensions of corporate strategies, the way policy frameworks develop and the changing dynamics of the organic movement, it is clear that the politics of organic food has undergone significant change. As corporations in the organic sector strive to meet growing consumer demand for 'clean, green and safe' foods, the technical qualities of the material end product have been institutionally privileged over many of the social and ecological values assigned to organic production processes.

Early practitioners of organic agriculture who believed in a principled way of producing food imagined that by keeping social and economic relations connected, an alternative means of food provision would emerge to provide people food that was produced in harmony with the environment. The overall goal of early practitioners and supporters of organic agriculture was to make it as ecologically sustainable as possible by omitting the use of synthetic chemicals to increase yields or kill pests. Equally significant to some practitioners was the aim of keeping *people* central to the way food production was organized. The health and well-being of producers, workers and consumers were just as high a priority as the health of the soil for some organic practitioners who believed in putting the process-based definition into practice. By spreading information about the social, economic and environmental benefits of organic farming to the general public, supporters of organic agriculture believed that once consumers were aware of how industrialized agriculture negatively affected their environment and their health, they would be motivated to engage in the promotion of the values associated with organic agriculture to change the social relations found in industrial agriculture. It was imagined that farmers and consumers engaging with each other through local organic food chains would be reconnected through personal, trust-based relationships that would foster a moral economy that rejected the practices of industrial agriculture.

One of the major political positions of the early movement was its refusal to include conventional agri-food corporations or national governments in the organization and management of organic value chains because of their support and promotion of industrialized agriculture and their lack of concern for soil health or environmental sustainability. Keeping out the main drivers of industrialization in the agri-food system was a fundamental aspect of keeping organic agriculture a viable challenge to the status quo. Originally, the goal was to establish a completely different food system that would not only challenge the industrial food system, but replace it. Organic practitioners and supporters hoped that once organic agriculture expanded and more people began to purchase organic foods, governments would have to shift support from industrial agriculture to organic agriculture. The hope was that 'cheap food policies', which subsidized exploitive conditions for farmers and workers within the agricultural sector, would be a thing of the past and that a more sustainable agri-food system that assured safe and healthy foods, working conditions and environmental practices would become the status quo.

Early organic practitioners staunchly opposed any type of political organization that reproduced the relationship between agribusiness and the government, that privileged economic efficiency over other substantive considerations relevant to food production. Independence from state and corporate interference motivated organic farmers to have decentralized and pluralistic forms of organization that allowed for a diversity of definitions of organic to exist, depending upon what members of private, localized farming groups decided organic should mean. Some groups emphasized social goals in their mandates, while other producers' groups put more emphasis on ecological sustainability. Having a flexible definition of organic allowed practitioners in different bioregions with various resources available to them the ability to practice sustainable agriculture without having too many restrictions or 'one size fits all' bureaucratic regulations imposed on them. Maintaining flexibility in farming styles while committing to sustainability allowed organic practitioners to remain relatively independent from centralized forms of decision-making.

However, the use of markets as a vehicle for social change and the absence of formal standards including the principles that were part of the process-based definition left the organic sector without signficant barriers to the entry of those with alternate definitions of organic. It is arguable that even if organic practitioners decided to codify the process-based definition of organic early on into the objectives of the movement, it would have been difficult to maintain ongoing distance from corporate actors while expanding the organic market. The product-based definition guiding production processes is now part of national regulations in both Canada and the US, and is also reflected in bilateral equivalency arrangements among major trading partners like Canada, the US and the EU. Though there are a number of benefits of applying competitive business principles in the organic sector – including providing more markets with organic products that cannot be produced locally and reducing the use of synthetic inputs in food production overall – the structural changes that conventional corporations have imposed upon the organic sector have transformed the politics of organic food. It is no longer a clear-cut alternative to the industrial food system.

The benefits and costs of the expansion of organic agriculture are nuanced and cannot simply be reduced to a 'good vs. bad' dichotomy. Instead, this book has attempted to show that changes in the organization and decision-making regarding organic agriculture and food has created a gray area between legally sanctioned organic practices and the principles that are commonly perceived as existing in all organic production systems. The changes have created an organic food sector that includes many different approaches, styles and values, which are not always clearly understood by consumers or represented on food labels. The consequence is the existence of multiple interpretations of organic, which may increase participation and diversity in practice but also serves to create a degree of ambiguity between what organic legally means, and what consumers think it means.

Despite the International Federation of Organic Agriculture Movements' (IFOAM) stated aim to keep social justice a part of the policy discourse of organic agriculture (IFOAM, 2002a), its limited political and economic influence in global trade policy-making has forced it to work from within the system and adjust its objectives accordingly. IFOAM has now included in its mandate ideas promoted by the World Trade Organization (WTO) to encourage states to decrease their 'disguised barriers to trade', whether they are rigorous environmental requirements or social standards included in national organic certification schemes. As organic agriculture became part of the global trade regime, its regulations were conditioned by the WTO's policies, which view the inclusion of Processes and Production Methods (PPMs) as potential areas of conflict between member states and disguised barriers to trade. '[Organic] standards are far more able to refer to prohibited inputs than to specify precise criteria for the assessment of whether producers and processors are acting in a manner that is "socially just" or "ecologically responsible"' (Rigby and Caceres, 2001:27). But for many who reject industrialized methods of food production or global trade's emphasis on the 'end product' regardless of process, the end product's value *is* derived from how it is produced and who produces it.

Not all aspects of the process-based definition of organic have been equally affected by expansion and transformation of the sector. For example, the spread of organic agriculture has environmental benefits, even though the product-based definition is now institutionalized into regulation and policy. The rising consumer demand for organic products has helped spread organic farming, resulting in the conversion of millions of hectares of farmland around the world from conventional to organic management, reducing the amount of synthetic chemicals used to produce food. As of 2012, there were 37.5 million hectares of certified organic land around the world, an increase from 11 million hectares in 1999 (Willer and Lernoud, 2014:23). While significant, expanding production is hardly exhaustive of the goals of the processed-based definition. Globalized organic food relies on the same transnational system of food transportation and distribution as conventionally produced food contributing to greenhouse gas emissions and climate change (McNeely and Scherr, 2001).

In 2011 there were five publicly traded TNCs that held significant decision-making power over the organic (and natural food and nutritional supplement) sector in North America: the Hain Celestial Group, UNFI, Whole Foods Market (WFM), SunOpta (now owned by UNF) and Green Mountain Coffee Roasters (now Kurig Green Mountain). These top five TNCs specializing in organic and natural foods collectively earned billions throughout 2014: SunOpta recorded $333.5 million (3Q, 2014); Kurig Green Mountain $1.02 billion; Hain Celestial had record net sales of $557.4 million (3Q, 2014); UNF $1.78 billion (3Q, 2014); and WFM $3.4 billion (3Q, 2014).[1] Their growing financial success largely stems from applying conventional business strategies to effectively absorb

---

[1]   All in USD. See 'Current News: UNFI Announces Third Quarter Fiscal 2014 Results', available at https://www.unfi.com/NewsAndEvents/Pages/UNFI AnnouncesThirdQuarter2014Results.aspx (accessed 19 January 2015); 'SunOpta Announces Fourth Quarter Results, 2014', available at http://investor. sunopta.com/releasedetail.cfm?ReleaseID=847693 (accessed 19 January 2014); 'Whole Foods Market reports third quarterly results', available at http://assets. wholefoodsmarket.com/www/company-info/investor-relations/financial-press-releases/2014/Q314-Financial.pdf (accessed 19 March 2015); 'Kurig Green Mountain: Quarterly results', http://investor.keuriggreenmountain.com/results. cfm (accessed 19 January 2015); 'PR Newswire, Hain Celestial announces record third quarter fiscal year 2014 results', available at http://www.pr newswire.com/news-releases/hain-celestial-announces-record-third-quarter-fiscal-year-2014-results-258441701.html (accessed 19 January 2015).

smaller organic and natural food firms (and the association of their labels with the process-based definition of organic) into their portfolios over the last 15 years. The organic sector is becoming more consolidated every year, which further complicates the politics of organic food and the way social activism contests corporate control in the food system. Organic food as a general category is no longer a clear alternative to industrialized food.

This book began by critically addressing the question Clunies-Ross (1990) posed 25 years ago: 'can organic farming remain true to its social movement roots while it economically expands?' The evolving cooperative relationship between corporations, national governments and global economic institutions over the last 50 years has significantly driven many of the changes observed in the organic movement. The creation of a system of standards and regulations for organic agriculture that reflects the structures and organization found in the conventional agricultural sector demonstrates how cooperative relationships between industry and government influence the content and structure of policy. It also demonstrates how largely informal practices traditionally outside of the mainstream decision-making structures are institutionalized. Regulation may have shifted decision-making control over standards 'from private to public' but that does not indicate that private interests no longer have influence over the content of standards and how they are applied. Corporate actors and organizations have had a crucial role, not only in standardizing productive processes but also in incorporating organic food into existing national and transnational policy frameworks. Consumer demand for organic products has been driven by fears over 'tainted food', biotechnology in the agri-food system and environmental degradation. Yet, this book shows that consumer demand and consumer activism are only part of the story of how the definition of organic has been organized and reorganized since the early twentieth century. Uncovering how decision-making regarding the organization of production processes has changed over time offers insight into how value-based movements transform in relation to the institutions they protect. It also shows how ideas – no matter how inconcievable they may sound at one point in time – can have real and lasting impacts on behaviors of governments, businesses and society in general.

Examining the changing politics of organic food reveals a number of things about how and why there is resistance to broader forces in the globalized economy. Taking a closer look at the changing structural dynamics of the organic sector shows that corporations have wielded

significant power in reshaping productive processes (that is, industrialization and transnationalization) and shaping policy directions of governments in Canada and the US (that is, product equivalence). Examining
the extent of corporatization in the organic sector also supports the claim
that as transnational interests grow in particular sectors, formally local
production processes are integrated into regional and global production
networks (Dicken, 1999; Gereffi and Korezeniewcz, 1994; Clapp and
Fuchs, 2009). Once corporate interests are privileged in the policy
process, public policy begins to reflect them over other interest groups
with fewer political and economic resources to draw on, or do not share
a unified position on what qualifies as appropriate policy action. This has
implications and lessons for other food movements beyond organic.

Liberal economic principles institutionalized in both the North American Free Trade Agreement NAFTA and the WTO have greatly influenced
policy formation in member countries in a number of economic sectors
(Hansenclever et al., 1997; Korten, 1995). National governments have
played important roles in liberalizing various sectors of their economies,
like agriculture, while cooperating in establishing global trade institutions
that strive to provide global economic stability signaling a shift towards
global governance (Cohen, 2007; Haas, 1990:59; Stoker 1998; Josling et
al., 2004). The development of organic agriculture policy at the national
and supranational levels reflects the shifting boundaries between the state
and civil society, as well as the influence that global economic institutions have on multi-level governance. The changing politics of organic
food speaks to the necessary compromises made by an alternative
approach to food production that was compelled to engage in the
normative behaviors found within the global economy, but also the
willingness to engage in formal regulatory frameworks that are, by
design, products of global governance.

In addition to the importance of the transformation of the definition of
organic to the organization of production processes, another central claim
of this book is that the shift from the private standards to public
regulation is a significant factor in organic agriculture's political institutionalization, as it shifted from a type of food production that was
privately controlled by disaggregated citizen-based groups to being
publicly regulated at multiple levels considering both the interests of
government and industry. Critical analysis of the structure and content of
policy for organic agriculture developed throughout this book contributes
to a better understanding of how liberal principles encourage the harmonization and convergence of regulations through global institutions.
Though states continue to have a degree of autonomy over decision-
making as demonstrated by the differences between national organic

standards in Canada and the US (discussed in Chapter 4), it is clear that once liberal market principles are incorporated into global political and economic institutions, significant policy divergence is made more difficult for states to maintain and pursue (Clarkson, 2004). The US however, has shown a greater degree of autonomy in designing its organic food policies because of its global economic power relative to Canada's. Canada's national organic standard was largely a result of trading partners' demands for a unified national standard and label for Canadian exports (Willer and Yuseffi, 2005). Thus, the development of the Canadian organic standard was stimulated by the desire of the Canadian government to keep global markets for Canadian organic food exports open, and to preserve its standing in the multilateral trading system, while assuring plans for bilateral agreements in the future. Canada's national standard reflects a significant degree of convergence towards the pre-existing standards of other policy regimes, specifically the US, but is also designed for the further integration of the Canadian organic sector into global governance structures by adhering to ISO and Codex standards covering food products.

By looking at the relationship between changing ideas, actors and the institutional contexts of organic food, we can see how these changes influence the strategies, structure and membership of social movements. The case of the organic movement shows how social movements based on market transactions as the main form of collective action are vulnerable to the influence of outside forces that can undermine the movement's original goals of social change. Early supporters like Sir Albert Howard and Wendel as well as skeptics of organic agricultural techniques from the 1940s until the 1970s believed that organic food would never become part of the mainstream since it was assumed that the structures and organization of organic production would not converge upon conventional production processes. The process-based definition of organic provided the movement with an explicit form of rejection of conventional economic and political structures, as well as actors (White, 1972; Merrill, 1976). As part of the process-based definition, the market for organic food products was constructed around keeping social and environmental relations integral parts of the production process. Originally, members of the organic movement viewed the market as a politically neutral institution and used it to create alternative agri-food networks; however, the early organic movement's reliance on market transactions and consumer activism, rather than policy-making, did little to protect it from those interested in capitalizing on the growing demand for organic products. As Bjarne Pedersen (2003:246) in his assessment of relying on the market to deliver social benefits notes:

when consumers are entrusted with the responsibility for continued development of sustainable food production, it is necessary to thoroughly examine the ability of the market to drive such development. In this respect, there may be some problems with a pure market model.

Treating the market as a 'place of opportunities' to challenge industrial forms of production is limited when the 'imperatives' of the market in a capitalist society are premised on exploitive social and environmental relations (Fridell, 2007:15). The market, as thought by some early organic practitioners who valued independence from the state, was viewed as a place where consumers could make their own choices and thus collectively influence production processes through market activities. Other early supporters of organic agriculture, who wanted to revolutionize the entire system of food production, believed that if consumers were given the right information about organic farming and the dangers of industrialized farming, more people would be willing to make the choice to support organic agriculture and it then would replace industrialized agriculture. Few early organic practitioners would have envisioned the 'two-tiered' agri-food system that exists today, with organic food available to those who have enough disposable income to afford it and 'cheap food' from the industrialized agri-food system for those who cannot (Hawaleshka et al., 2004:22). The dualistic model of food availability was not the original intention of the organic movement but is the reality of the current food system.

Tracing the rise of conventional business strategies in the organic sector has revealed insights into how other contemporary consumer-based movements may evolve in the context of increasing corporate consolidation in agriculture. For example, criticisms have surfaced regarding the fair trade network, and its use of conventional transnational supply chains, suggesting that its activist elements have been coopted by corporate interests seeking to capitalize on its quality assurance labels with a premium price tag (Rice, 2001; Ransom, 2005). The case of the evolving organic movement shows that the market is not a politically neutral institution. Thus, excluding the state as a focus of lobbying and pressure to achieve social change, and instead focusing on achieving change through the market, may make movements more susceptible to competitive corporate behaviors traditionally rejected by those seeking an alternative to the status quo, especially when priority is given to certain aspects (for example, restricted use of off-farm inputs) that define the goals of the movement over others (for example, labor practices). Movements that are vulnerable to cooptation because of their orientation towards market activities must also target political structures and include

political goals (policy change) in their instruments of social change if
they are to mount a resistance to the status quo. Though professionalization and institutionalization of social movements present challenges to
keeping a movement responsive to its grassroots principles and members
(Tilly, 2004; della Porta and Tarrow, 2005), there is something to be said
for the pre-emptive institutionalization of founding principles and norms.
Doing so in the early stages of a social movement's development can
erect barriers of entry to new members who might seek to change its
objectives (Coy and Hedeen, 2005). Legally enforcing the inclusion of
process as a fundamental part of the organic label would have presented
greater barriers to the emergence of the product-based definition. Actors
wishing to participate in a market for 'synthetic input free' products may
have had to choose another label, such as 'natural' (Fromartz, 2006). The
current natural versus organic labeling debate stems from ongoing
confusion and misinformation about what organic actually refers to and
what people think it means (Abrams, Meyers and Irani, 2010).
The co-existence of dual interpretations of organic contribute to this
confusion.

ORGANIC AND CONTESTING GLOBALIZATION

Although the power of economic globalization is sometimes viewed as
inevitable and unstoppable, unique forms of social resistance continue to
mount challenges against it (Lipschutz, 1992; Ayres, 1998; Guidry, 2000;
Clapp and Fuchs, 2009; Andree et al., 2014). Though there is a plethora
of dynamic and progressive food movements actively challenging the
norms and behaviors of the globalized food system, the focus here is on
the current state of organic food politics and its ability to challenge
industrial agriculture. This book has not claimed that business principles
or market participation in the organic sector are necessarily detrimental to
keeping the process-based definition in practice per se. Rather, it suggests
that the type of corporate behavior found in the conventional agri-food
system privileged in national policies and trade agreements is the force
that fundamentally changed the organization of the organic sector and the
role the process-based definition of organic has in food production. Thus,
as a United States Department of Agriculture study on the growing
market for organic products in the US claims, 'if consumers do not
demand that organic products be environmentally and socially sustainable
there is a danger that these aspects may be forgone in the production
process all together' (Dimitri and Richman, 2000:2). Some organic
producers, in an effort to resist the shifting definition of organic and the

erosion of the process-based definition of organic, have opted to use other words like 'authentic' to describe the qualities of their 'organically grown' products (Merrigan, 2003:280). Others have protested by avoiding national certification schemes all together, instead relying on trust-based relationships established through short value chains that once populated early organic production systems (Logsdon, 1993; Seiff, 2005). More recently in the US, a movement is underfoot to create a 'certified natural' labeling scheme. Largely in response to the 'government takeover' of the organic program in the US, a group of farmers in New York State in 2002 created the 'Certified Naturally Grown' label program. It now includes more that 700 farms across 47 states (Esch, 2013). Foods carrying the label are produced without synthetic chemicals, are inspected by other farmers and distributed in local markets. The food products are grown organically, but do not participate in the organic certification program. Some farmers have 'philosophical objections to joining a monolithic government-run program that also certifies huge operations that ship produce across the country' (Esch, 2013). Yet, similar to the regulatory structure governing 'certified' organic producers, there is no mention of labor practices or suggestion of linking 'naturally grown' production practices with social change. Certified Naturally Grown may have more in common with the institutionalized 'certified organic' than practitioners realize. Clearly, resistance to the dominant product-based definition of organic is still active, though is taking a different shape than it has in the past.

Other businesses emerging from the organic movement that are now a part of the organic advocacy network have resisted the corporatizing forces of the globalization and have tried to remain true to the process-based definition of organic. US-based Swanton Berry Farm and Canadian company Nature's Path are two organic/natural food businesses that continue to mount a challenge to the product-based definition of organic. Swanton Berry Farm is located in California and was established in 1983 covering almost 200 acres near Santa Cruz. Cofounder Jim Cochran practices polyculture and crop rotation, focusing most of the farm's production on organic strawberries. In 1987, Swanton Berry Farm became certified organic along with other farms across California. Cochran and Swanton Farm were recognized by the US Environmental Protection Agency for the pioneering efforts to farming strawberries without the use of methyl bromide, a major ozone depleter and a health hazard for farm workers. In 1998, Swanton was the first organic farm to sign a contract with the United Farm Workers American Federation of Labor/Congress of Industrial Organizations (AFL/CIO), making it a fully unionized certified organic enterprise. Decision-making is relatively disaggregated and employees take responsibility for various aspects of the food production. It is not only unionized, but employees are also able to purchase stock bonuses and are provided with health care insurance and access to low-cost housing. The position of Swanton Berry Farm toward fair labor practices is that 'the existence of a union contract formalizes our commitment to the human side of the farming equation, much as the process of organic certification formalizes our commitment to a set of farming practices' (Swanton Berry Farm, 2012). Swanton Berry Farm is an example of how organic practitioners who believe in a more process-based definition of organic are putting it into practice. Cochran, nine other co-owners and Swanton's employees are committed not only to the environmental benefits of organic techniques, but also the principles of social sustainability and fair labor practices championed by members of the process-based definition of organic.

British Columbia-based Nature's Path is another example of a small company that has managed to remain true to the core values of the process-based definition of organic even as it has expanded. Nature's Path produces a variety of certified organic processed foods such as breakfast cereals and granola bars. Nature's Path has remained independent since its establishment by Arran and Ratana Stephens in 1985. Arran Stephens comes from a farming background and is still the CEO of Nature's Path. The company was created out of the LifeStream natural foods line that Stephens began in 1977. Stephens has fought to remain independent, and after Kraft/Philip Morris acquired LifeStream in 1981, Stephens bought it back in 1995 when Nature's Path achieved enough financial success to allow him to do so (Nature's Path, 2013). Nature's Path has facilities in British Columbia and Washington state, sourcing most of its ingredients as locally as possible. Although Nature's Path is the largest distributor of organic cereals in North America, Stephens has remained an active member of civil society, joining forces with the Council of Canadians to speak out against GMOs in the agri-food system and the corporate consolidation of the agri-food sector in North America (Council of Canadians, 2005). Although the product-based definition has gained momentum as consumer demand increases for competitively priced, organically grown products, both Swanton Berry Farm and Nature's Path are examples of how it is possible for businesses to be functional and profitable in the organic sector while including social and environmental commitments in business activities.

In addition to businesses fighting to maintain a place for the process-based definition of organic in the corporatized organic sector, some members of civil society have adjusted their tactics to promoting local agri-food chains as the most socially, environmentally and economically sustainable way to produce food while challenging the practices in the industrial agri-food system (Steele, 1995; Lang and Heasman, 2005; McMichael, 2003). Organic is still promoted because of its contribution to reducing the use of synthetic inputs in food production, but local production networks are argued by some to be the best way to achieve food sustainability and security in order to fight corporatization of the agri-food system. Local NGOs with a focus on food issues have actively campaigned on a platform that links a number of the issues. Food policy networks work to develop policy options that link food sustainability and security by promoting the relocalization of agri-food value chains (Friedmann, 2007:392). Some members of food policy networks, like Toronto-based FoodShare and Vancouver-based FarmFolk/CityFolk (FFCF), work with the community by hosting public forums, supporting farmer's markets and distributing information to promote local systems of production, which are seen as the best means of achieving food security and sustainability over practicing formal organic production methods.

FFCF is a firm supporter of organic agriculture, but believes that mounting a more diverse social challenge to the industrialized agri-food system is by far the best way to reconnect the social, environmental and economic spheres of agricultural production (FFCF, 2013). Localized agri-food value chains that include re-establishing trust-based relationships and eating in season are possible ways to protest against the industrialized agri-food system. One of the most important aspects of FFCF work is that it goes beyond merely promoting the consumption of sustainably produced food to also supporting producers, and lobbies for policy change at the local and provincial level, along with other NGOs such as Canadian Organic Growers. It is essential that supporters present viable policy options to governments which insist on community members having an active role in policy design. Above all, social change cannot hinge on 'conscious consumption' as the only means of achieving social change in the agri-food sector, since this is a highly undemocratic form of representation that excludes members of society who cannot participate in ethical consumption which is largely contingent on economic status.

Groups like the National Farmers Union (NFU), functioning in both Canada and the US, support sustainable agriculture and organic agriculture (although not exclusively), and have presented policy documents to various levels of government outlining the promise of sustainable agriculture as the only option that will support farmers and the environment in the future (NFU, 2013). Localizing agri-food systems is now viewed as one strategy to reconnect the social and economic relations that are lost through industrializing processes, while reducing food production's impact on the environment (Friedmann, 2007). However, local food chains are not a panacea for the problems that exist in the food system. In some cases, the emphasis on 'local' above other aspects of process has overshadowed other equally important goals of a sustainable food system, such as fair labor practices and reducing dependence on fossil fuels in production and transportation. The key to a sustainable agri-food system that guarantees the security and safety of the food supply in addition to maintaining communities is, as J. Ann Tickner suggests, 'changing our relationship with nature … [Only then] can real security, for both our natural environment and its human inhabitants be assured' (Tickner, 1993:66).

By showing how organic production processes were institutionalized and transnationalized, and how the notion of organic changed over time, this book has discussed how some of the most important aspects of the process-based definition of organic, such as social and ecological goods, have lost their standing as vital elements of what it technically means for a food to be organic. One of the major lessons we can learn from examining the changing politics of organic food is that relying on the market as the primary means for social change puts substantive goals at risk when liberal market principles govern how the economy is organized and regulated. There are various ways of contesting and resisting economic globalization from the 'bottom up' by promoting forms of sustainability through engagement in the political process. Moving beyond the limitations of the current formalized definition of 'organic' and working towards cross-issue linkages that emphasize the importance of people in the food system are increasingly necessary to develop potential solutions to the global economic, social and environmental challenges we all face.

Appendix

CORPORATE ACQUISITIONS, MERGERS, BRAND INTRODUCTION AND PARTIAL EQUITY IN THE ORGANIC AND NATURAL FOOD SECTOR IN CANADA AND THE US

(This is not an exhaustive list.)

2014: Acquisitions – 5[1]

- Hain Celestial acquires Rudi's Organic Bakery
- General Mills acquires Annie's Inc.
- Hillshire Farms acquires Van's Natural Foods
- Post Holdings Inc. acquires Michael Foods Inc.
- Treehouse Foods acquires Protenergy Natural Foods

2013: Acquisitions – 2; partial equity – 1

- Danone gains partial equity in Happy Family
- Campbell's Soup Co. acquires Plum Organics
- Hain Celestial acquires Ella's Kitchen

2012: Acquisitions – 12

- JAB/D.E. Master Blenders (formally Sara Lee) acquires Tea Forte and Peet's Coffee and Tea
- Miller-Coors acquires Crispin and Fox Barrel
- J&J Snack Foods acquires Kim & Scott's
- Post Foods (spin-off from Ralcorp) acquires Erewhon and New Morning

---

[1] Sources for the information in this appendix: Howard, 2014, 2006, 2007, 2008, 2009a; Sligh and Christman, 2003; Glover 2005; Draffan, 2004; Organic Monitor, 2005a, 2005b, 2006; SunOpta Annual Report, 2003, Hain Celestial 2006b, 2013; OCA, 2007; Capstone Partners, 2011, 2013.

- TreeHouse Foods acquires Naturally Fresh and Strum Foods
- Campbell's soup acquires Bolthouse Farms
- Hain Celestial acquires BluePrint
- General Mills acquires Food Should Taste Good

2011: Acquisitions – 9

- Hain Celestial acquires Danival SAS (FR)
- Nestlé acquires Sweet Leaf Tea
- Delta Partners acquires Royal Wessanen
- Perdue Farms acquires Coleman Natural Foods
- Nutrition & Sante Iberia acquires Natursoy
- Cooper Tea Company acquires 3rd Street Chai
- TSG Consumer Partners acquires Stumptown Coffee Roasters
- Kensington Energy acquires Natures Prime Organic Foods
- Hillshire Brands acquires Aidell's Sausage

2010: Acquisitions – 9; strategic alliances – 1

- Mondelez (spinoff of Kraft) acquires Green & Black's
- United National Foods acquires SunOpta
- Solbar USA acquires SPECIALTY Protein Producers
- Avenir Enterprises Gestion acquires Balarama
- Meyer Natural Foods acquires Howard Venture
- Hearthside Foods acquires Golden Temple, Peace Cereal and Williamette Valley Granola
- John B. Sanfilipo & Son acquires Orchard Valley Harvest
- Cargill signs a joint marketing agreement with Meyer Natural Foods

2009: Acquisitions – 2

- Foster Farmers acquires Humbolt Creamery
- TreeHouse Foods acquires Strum Foods

2008: Acquisitions – 1

- Hain Celestial acquires MaraNatha

2007: Acquisitions – 11; partial equity – 2

- WholeFoods Markets acquires Wild Oats Markets
- United Natural Foods acquires Millbrook
- Tree of Life aquires Organica
- Hain Celestial acquires Tofu Town
- Kellogg acquires Bear Naked and Wholesome & Hearty
- SIG Strategic Investments acquires distributor US Mills
- Campbell's Soup acquires Wolfgang Puck
- Clearly Canadian acquires Crofter's Organics, distributor
- DMR Foods and My Organic Baby
- Nestlé gains partial equity Tribe Mediterranean Foods
- Coca Cola gains partial equity in Honest Tea

2006: Acquisitions – 4; brand introduction – 7

- SunOpta acquires Purity Life Health Products
- Hersey Foods acquires Dagoba
- Pepsi acquires Naked Juice
- Rich Products Corp. acquires French Meadow
- ConAgra introduces PAM Organic
- Kellogg introduces Keebler Organic and Kellogg's Organic
- nSpired Natural Foods introduces O'Coco's
- Sobeys introduces Compliments Organic
- Walmart introduces Parent's Choice and Great Value

2005: Acquisitions – 6; brand introduction – 6; partial equity – 1; strategic alliances – 1

- Cadbury-Schweppes acquires Green & Black's
- Charterhouse Inc. acquires Rudi's Organic Bakery, and significant equity in The Vermont Bread Co.
- Hain Celestial acquires Spectrum Organic Products
- Monsanto acquires Seminis (conventional and organic seeds)
- United Natural Foods acquires Roots & Fruits
- Homegrown Naturals acquires Annie's Naturals
- ConAgra introduces Hunt's Organic, Orville Redenbacher's Organic
- Sainsbury (UK) launches So Organic
- UniLever introduces Ragu Organic
- Safeway introduces O Organics and Safeway Select
- Hain Celestial forges strategic alliance with Yeo Hiap Seng (Asia)

2004: Acquisitions – 8

- Cadbury-Schweppes acquires Nantucket Nectar
- Dean Foods acquires Horizon Dairy
- Hain Celestial acquires JASON products, Harry's Snacks and Kineret[2]
- HJ Heinz acquires Linda McCartney Vegetarian Foods[3]
- Kraft acquires Balance Bars
- SunOpta acquires full equity of Organic Ingredients Inc. (dist.)
- Whole Foods Markets acquires Fresh & Wild (UK) (supermarket)

2003: Acquisitions – 10; brand introduction – 5; strategic alliances – 1

- Clement Pappas acquires Crofter's Organic Juices
- Groupe Danone acquires Brown Cow
- Hain Celestial acquires Acirca and Walnut Acres and Grains Noirs (BEL)
- Horizon acquires Rachel's Organic (UK)[4]
- Kraft acquires Back to Nature
- Nestlé acquires Poland Spring Water[5]
- SunOpta acquires Kettle Valley and ProOrganics (dist.)
- Campbell's introduces Campbell's Organic
- PepsiCo introduces Tostitos Organic
- UniLever introduces Ben and Jerry's Organic
- Wholefoods introduces Whole Kids, Authentic Food Artisan
- Hain Celestial forges strategic alliance with Cargill

2002: Acquisitions – 9; partial equity – 1; brand introduction – 3; strategic alliances – 1

- American Capital acquires Coleman Natural Products
- Booth Creek acquires Petaluma Poultry
- Cadbury-Schweppes acquires Hanson Natural
- Dean Foods acquires White Wave/Silk
- Hain Celestial acquires Imagine Foods, Rice Dream and Soy-Dream

---

[2] Listed in Brand names at Hain Celestial Website; date of acquisition unknown.
[3] Listed on HJ Heinz website; date of acquisition unknown.
[4] As listed in Sligh and Christman, 2003.
[5] See Glover, 2005.

- SunOpta acquires Wild West and Simply Organic
- Solera gains significant equity in Homegrown Naturals
- HJ Heinz introduces Heinz Organic
- Whole Foods Market introduces '365 Everyday Organic Value'
- SunOpta introduces MU
- Cargill forges strategic alliance with French Meadow

2001: Acquisitions – 6; partial equity – 1; brand introduction – 3

- Coca-Cola acquires Odwalla Organics
- Hain Celestial acquires Friti De Bosco, Millina's Finest, Mountain Sun, Yves Veggie Cuisine (CAN), Shari Ann's and Lima (BEL)
- Groupe Danone acquires partial equity in Stoney Field Farms
- Dole introduces Dole Organic
- Loblaws introduces President's Choice (PC) Organics
- Tyson introduces Nature's Farms

2000: Acquisitions – 14

- Kraft acquires Boca Burger
- Hain Food Group acquires Celestial Seasonings
- ConAgra acquires Fakin' Bakin, Light Life, Foney Baloney, Gimme Lean, Smart Dogs, Smart Menu Strips and International Home Foods
- Kellogg acquires Kashi
- Whole Foods Market acquires Food 4 Thought Natural Food Market and Deli
- UniLever acquires Best Foods and Ben & Jerry's
- Homegrown Naturals acquires Fantastic Foods (renamed Fantastic World Foods)

1999: Acquisitions – 12; brand introduction – 1

- HJ Heinz invests $100 million in Hain Food Group (20%)
- Dean Foods acquires Alta Dena and Organic Cow of Vermont
- Hain Food Group acquires BreadStop, Casbah, Earth's Best, Health Valley and West Soy
- General Mills acquires Cascadian Farms, and Muir Glen
- Kellogg acquires Morning Star Farms and Worthington Foods
- Tanimura and Antle acquires Earth Bound Farms
- Muir Glen introduces Sunrise Organic

1998: Acquisitions – 6

- Hain Food Group acquires Terra Chips, Deboles, Garden of Eatin and Arrowhead Mills
- HJ Heinz acquires Nile Spice
- United Natural Foods acquires Albert's Organics

1997: Acquisitions – 5

- Hain Food Group acquires Bearitos, Little Bear and Westbrae
- M&M/Mars acquires Seeds of Change
- General Mills acquires Small Planet Foods

1996: Acquisitions – 1

- Wild Oats acquires Capers Community Markets (BC)

1995: Merger – 1; brand introduction – 1

- Cornucopia Natural Foods and Mountain People's Warehouse merge to create United Natural Foods
- General Mills introduces Gold Medal Organic

1994: Acquisitions –1

- Smuckers acquires After the Fall

1989: Acquisitions – 1

- Smuckers acquires Santa Cruz Organic

1987: Acquisitions – 1

- Nestlé acquires Arrowhead Water

1985: Acquisitions – 1

- Cornucopia Natural Foods acquires Earthly Organics

1984: Acquisitions – 1; mergers – 1

- Smuckers acquires RW Knuden
- Earthbound Farms merges with Mission Ranches

1981: Acquisitions – 1

- Kraft/Philip Morris acquires LifeStream

1980: Merger – 1

- Safer Way Natural Foods & Clarksville Natural Grocer merge to form Whole Foods Market in Austin, Texas

Sources: Howard, 2014, 2006, 2007, 2008, 2009a; Sligh and Christman, 2003; Glover 2005; Draffan, 2004; Organic Monitor, 2005a, 2005b, 2006; SunOpta Annual Report, 2003, Hain Celestial 2006b, 2013; OCA, 2007; Capstone Partners, 2011, 2013.

Bibliography

Abaidoo, Samuel and Harley Dickinson (2002), 'Alternative and Conventional Agricultural Paradigms: Evidence From Farming in Southwest Saskatchewan', *Rural Sociology,* **67**(1), 114–31.

Abrams, Katie, Courtney A. Meyers and Tracy A. Irani (2010), 'Naturally Confused: Consumers' Perceptions of All-natural and Organic Pork Products', *Agriculture and Human Values,* **27**(3), 365–74.

ACNielsen Canada (2009), 'Grocery Label Scan Study, February 2009', available at http://www4.agr.gc.ca/AAFC-AAC/displayafficher.do?id= 1285870839451&lang=eng (accessed 7 October 2012).

Adamchak, Raoul (2008), 'The Farm', in Ronald, Pamela and Raoul Adamchak, *Tomorrow's Table: Organic Farming, Genetics and the Future of Food*, New York: Oxford University Press, pp. 13–40.

Agriculture and Agri-food Canada (AAFC) (2014), 'Organic Agriculture 32/20', available at http://www.tpsgc-pwgsc.gc.ca/ongc-cgsb/programme-program/normes-standards/comm/32-20-agriculture-eng.html (accessed 13 January 2014).

AAFC (2009), 'Canada's Organic Industry at a Glance 2009', available at http://www4.agr.gc.ca/AAFC-AAC/displayafficher.do?id=1276292934 938&lang=eng (accessed 13 September 2012).

Ahn, Christine, Melissa Moore and Nick Parker (2004), 'Migrant Farmworkers: America's New Plantation', *Backgrounder,* **10**(2), Institute for Food and Development Policy.

Akerlof, George A. (1970), 'The Market for "Lemons": Quality Uncertainty and the Market Mechanism', *The Quarterly Journal of Economics,* **84**(3), 488–500.

Allen, Linda (2012), 'The North American Agreement on Environmental Cooperation: Has It Fulfilled Its Promises and Potential? An Empirical Study of Policy', *Colorado International Journal of Environmental Law and Policy*, **23**(1), 123–99.

Allen, Patricia and Martin Kovach (2000), 'The Capitalist Composition of Organic: The Potential of Markets in Fulfilling the Promise of Organic Agriculture', *Agriculture and Human Values,* **17,** 221–32.

Allen, Patricia and Carolyn Sachs (1991), 'The Social Side of Sustainability: Class, Gender and Race', *Science as Culture,* **12,** 569–90.

Alteri, Miguel A. and Peter M. Rosset (1997), 'Agroecology versus Input Substitution: A Fundamental Contradiction of Sustainable Agriculture', *Society and Natural Resources*, **10**(6), 283–95.

Amaditz, K.C. (1997), 'The Organic Foods Production Act of 1990 and its Impending Regulations: A Big Fat Zero for Organic Food?', *Food Drug Law Journal*, **52**(4), 537–59.

Andree, Peter, et al. (eds) (2014), *Globalization and Food Sovereignty: Global and Local Change in the New Politics of Food*, Toronto: University of Toronto Press.

Andree, Peter (2011), 'Civil Society and Political Economy of GMO Failures in Canada: A Neo-Gramscian Analysis', *Environmental Politics*, **20**(2), 173–91.

Armstrong, David (1981), *Trumpet to Arms: Alternative Media in America*, New York: South End Press.

Associated Press (2007), 'Whole Foods CEO "bluntly" outlines merger bid, *MSNBC*,' available at http://www.msnbc.msn.com/id/19315379/ (accessed 12 July 2010).

Atkins, Peter and Ian Bowler (2001), *Food in Society: Economy, Culture, and Geography*, New York: Oxford University Press.

A.W. Page Society (2009), 'Whole Foods/Wild Oats Merger: Sowing the Seeds for Market Growth', available at http://www.awpagesociety.com/wp-content/uploads/2011/09/WholeFoods_CaseStudy.pdf (accessed 17 September 2012).

Ayres, Jeffrey M. (1998), *Defying Conventional Wisdom: Political Movements and Popular Contention Against NAFTA*, Toronto: University of Toronto Press.

Baker, Brian (2004), 'Introduction: Brief History of Organic Farming and the National Organic Program', *Organic Farming Compliance Handbook: A Resource Guide for Western Region Agricultural Professionals*, Palo Alto: University of California Davis.

Balfour, E.B. (1943), *The Living Soil: Evidence of the Importance to Human Health of Soil Vitality, with Special Reference to National Planning*, London: Faber and Faber Ltd.

Barrett, H.R. et al. (2002), 'Organic Certification and the UK Market: Organic Imports from Developing Countries', *Food Policy*, **27**(4), 301–18.

Belasco, Warren L. (1989), *Appetite for Change: How the Counterculture Took on the Food Industry 1966–1988*, New York: Pantheon Books.

Bell, David and Gill Valentine (1997), *Consuming Geographies: We Are What We Eat*, London: Routledge.

Bellon, Stephane and Servane Penvern (eds) (2014), *Organic Farming, Prototype for Sustainable Agricultures*, New York: Springer.

Bennett, Colin (1989), 'Review Article: What is Policy Convergence and What Causes it?', *British Journal of Political Science*, **21**, 215–33.

Bennett, W. Lance (2005), 'Social Movements Beyond Borders: Understanding Two Eras of Transnational Activism', in Donatella della Porta and Sidney Tarrow (eds), *Transnational Protest and Global Activism*, Toronto: Rowman & Littlefield Publishers.

Bentley, Stephen and Ravanna Barker (2005), *Fighting Global Warming at the Farmer's Market: The Role of Local Food Systems in Reducing Greenhouse Gases*, Toronto: FoodShare Research in Action Report.

Berry, Wendell (1977), *The Unsettling of America: Culture and Agriculture*, San Francisco: Sierra Club Books.

Berry, Wendell (1976), 'Where Cities and Farms Come Together', in Richard Merrill (ed.), *Radical Agriculture*, New York: New York University Press, pp. 14–25.

Berry, Wendell (1970), 'Think Little', *The Whole Earth Catalog*, September, 3–5.

Berry, Wendell (1969), 'An Agricultural Testament', *The Whole Earth Catalog*, Spring, 33.

Bjorkhaug, Hilde (2004), *Is there a 'Feminine Principle' of Farming – and is Organic Farming a Way of Expressing It?*, paper no 5/04 presented at 'Globalisation, Risks and Resistance' World Congress of Rural Sociology, Norway. 25–30 July.

Blowfield, Mick (2001), 'Ethical Trade and Organic Agriculture', *Tropical Agriculture Association Newsletter*, March, 22–6.

Blum, Andrea (2006), 'Organic Farming's Labor Problem', *Common Ground*, February, available at http://www.columbia.org/pdf_files/ cainstituteforruralstudies.pdf (accessed 12 September 2012).

Bonanno, Alessandro et al. (eds) (1994), *From Columbus to ConAgra: The Globalization of Agriculture and Food*, Lawrence: University Press of Kansas.

Bookchin, Murray (1976), 'Radical Agriculture', in Richard Merrill (ed.), *Radical Agriculture*, New York: Harper Collins Books, pp. 3–13.

Borenstein, Seth (2005), 'Administration Kept Mum about Unapproved Modified Corn Sold', *Knight Ridder Newspapers* (March 22) available at http://www.mcclatchydc.com/2005/03/22/11314/administration-kept-mum-about.html (accessed 23 November 2013).

Bostrom, Magnus and Mikael Klintman (2006), 'State-centred versus Nonstate-driven Organic Food Standardization: A Comparison of US and Sweden', *Agriculture and Human Values,* **23**,163–80.

Bowen, Diane (2002), 'OECD Workshop on Organic Agriculture: International Harmonization of Organic Standards and Guarantee Systems,'

Bonn: IFOAM, available at https://web.archive.org/web/20050204181
713/http://www.ifoam.org/orgagri/oecd_harmonization_paper.html (accessed 2 July 2012).

Bove, Jose and Francois Dafour (2001), *The World is Not For Sale*,
London: Verso.

Brand, Karl-Werner (1990), 'Cyclical Aspects of New Social Movements: Waves of Cultural Criticism and Mobilization Cycles of New
Middle-class Radicalism', in Russell J. Dalton and Manfred Kuechler
(eds), *Challenging the Political Order: New Social Movements in
Western Democracies*, Cambridge: Polity Press, pp. 23–42.

British Columbia, Government of (1993), 'Agri-Food Choice and Quality
Act: Certification Regulation, B.C. Reg. 200/93', available at http://
www.bclaws.ca/EPLibraries/bclaws_new/document/ID/freeside/10_200
_93 (accessed 4 December 2012).

Broome, J.C. (2012), *Pest Notes: Bordeaux Mixture*, UC ANR Publication 7481. Davis: University of California, available at http://www.
ipm.ucdavis.edu/PMG/PESTNOTES/pn7481.html (accessed 17 January
2013).

Brownstone, Sydney (2014) 'Americans will pay more for organic, but
they also have no idea what it means: The food marketers have
won, FastCoExist', available at http://www.fastcoexist.com/3038415/
americans-will-pay-more-for-organic-but-they-also-have-no-idea-what-
organic-means (accessed 30 December 2014).

Buck, Daniel et al. (1997), 'From Farm to Table: The Organic Vegetable
Commodity Chain of Northern California', *Sociologica Ruralis*, **37**(1),
3–20.

Bullock, David S. et al. (2000), *The Economics of Non-GMO Segregation
and Identity Preservation*, available at http://ageconsearch.umn.edu/
bitstream/21845/1/sp00bu03.pdf (accessed 5 May 2013).

Burros, Marian (2007), 'Is Whole Foods Straying From Its Roots?', *New
York Times*, 28 February.

Burros, Marian (1989), 'A Growing Harvest Of Organic Produce', *New
York Times*, 29 March.

Burros, Marian (1987), 'The Fresh Appeal of Foods Grown Organically',
New York Times, 28 January.

Burt, S.L. (2000), 'The Strategic Role of Retail Brands in British
Grocery Retailing', *European Journal of Marketing,* **34**(8), 875–90.

Burt, S.L. and L. Sparks (2002), 'Corporate Branding, Retailing, and
Retail Internationalization', *Corporate Reputation Review,* **5**(2/3), 194–
212.

Business Week (1980), 'Rodale Reaches Out for the Mainstream',
Business Week, 27 October, pp. 85–8.

Buttel, Fredrick (1997), 'Some Observations on Agri-food Change and the Future of Agricultural Sustainability Movements', in David Goodman and Michael Watts (eds), *Globalising Food: Agrarian Questions and Global Restructuring*, New York: Routledge, pp. 344–67.

Buttel, Fredrick and Peter LaRamee (1991), 'The Disappearing Middle: A Sociological Perspective', in W. Friedland et al. (eds), *Towards a New Political Economy of Agriculture*, Boulder: Westview Press, pp. 173–88.

Cacek, Terry and Linda L. Langner (1986), 'The Economic Implications of Organic Farming', *American Journal of Alternative Agriculture*, **1**(1), 25–9.

California Certified Organic Farmers (CCOF) (1974), 'CCOF Standards of Farm Certification', *The California Certified Organic Farm Newspaper*, Spring 1974.

Campbell, B.L., S. Mhlanga and I. Lesschaeve (2013), 'Perception Versus Reality: Canadian Consumer Views of Local and Organic', *Canadian Journal of Agricultural Economics/Revue canadienne d'agroeconomie*, **61**, 531–58.

Campbell, Hugh and Christopher Rosin (2011), 'After the "Organic Industrial Complex" an Ontological Expedition through Commercial Organic Agriculture in New Zealand', *Journal of Rural Studies*, **27**, 350–61.

Canada (government of), the Government of United Mexican States and the Government of United States of America (1994), *The North American Free Trade Agreement text: final version including supplemental agreements*, Chicago: Commerce Clearing House.

Canadian Biotechnology Action Network (CBAN) (2012), 'GE Free Zones', available at http://www.cban.ca/Resources/Topics/GE-Free-Zones (accessed 2 November 2012).

Canadian Broadcasting Corporation (CBC) (2006a), 'Carrot juice recalled due to botulism concerns; CFIA', available at http://www.cbc.ca/consumer/story/2006/10/02/carrot-recall.html (accessed 1 August 2012).

CBC (2006b), 'E. coli Tainted Spinach Sickens 109 in the U.S.', available at http://www.cbc.ca/world/story/2006/09/18/spinach.html (accessed 18 September 2012).

Canadian Food Inspection Agency (CFIA) (2014), 'Food Recall Warnings: High risk,' available at http://www.inspection.gc.ca/about-the-cfia/newsroom/food-recall-warnings/eng/1299076382077/1299076493846 (accessed 19 January 2015).

CFIA (2012), 'Canada–US Organic Equivalence Arrangement – Overview', available at http://www.inspection.gc.ca/food/organic-products/

equivalence-arrangements/us-overview/eng/1328068925158/132806901
2553 (accessed 17 September 2012).

CFIA (2003), '2003 Guide to Food Labelling and Advertising; Chapter
4', available at www.inspection.gc.ca/english/fssa/labeti/guide/ch4ae.
shtml (accessed 30 March 2010).

Canadian General Standards Board–National Standard of Canada
(CGSB) (2006), 'Organic Production Systems: General Principles and
Management Standards', available at http://www.pwgsc.gc.ca/cgsb/on_
the_net/organic/032_0310_2006-e.pdf (accessed 4 October 2012).

CGSB (1999), 'Organic Agriculture: CAN/CGSB-32.310-99', available
at www.pwgsc.gc.ca/cgsb/032_310/32.310epat.pdf (accessed 10 February 2013).

Canadian Organic Growers (COG) (2014), 'About Canadian Organic
Growers', available at http://www.cog.ca/about/about_canadian_organic_
growers/ (accessed 15 June 2014).

COG (2012), *Scaling Up Organically*, Ottawa: COG publications.

COG (2011), 'Policy Development', available at http://www.cog.ca/our-
work/promoting-organic-farming/policy-development/ (accessed 30 July
2012).

COG and Laura Telford (2006), 'Letter to Dr. Bashir Manji, Re:
Canadian Organic Growers: Comments on Organic Products Regu-
lations/Règlement sur les produits biologiques', *Canada Gazette*, **140**
(35), available at http://www.cog.ca/documents/COGResponsetoOrganic
ProductsRegulation_002.pdf (accessed 14 December 2013).

Capstone Partners, Investment Banking Advisors (2013), 'Natural and
Organic Food & Beverages: Coverage Report, Q1 2013', available at
http://www.capstonellc.com/sites/default/files/Capstone%20Natural%20
Products_Q1%202013.pdf (accessed 8 January 2014).

Capstone Partners, Investment Banking Advisors (2012), 'Natural and
Organic Food & Beverages: Coverage Report, Q4 2011', available at
http://www.capstonellc.com/research/industryreports/Natural%20and%
20Organic%20Food%20and%20Beverage%20Coverage%20Report%20
2011.pdf (accessed 26 November 2012).

Carson, Rachel (1962), *Silent Spring*, Greenwich: Fawcett Crest.

Carter, Luther J. (1980), 'Organic Farming Becomes "Legitimate"',
Science, **209**, 254–6.

Castairs, Catherine (2006), 'Health Food Craze Sign of Growing Mistrust,' available at http://www.uoguelph.ca/news/2006/05/health_food_
cra.html (accessed 5 May 2012).

Caswell, Julie (1997), 'Uses of Food Labelling Regulations', *OECD
Working Papers 5, no. 100*, Paris: OECD Publications.

Center for Disease Control and Prevention (CDC) (2013), 'CDC Features: Trends in Foodborne Illness in the United States, 2012', available at http://www.cdc.gov/Features/dsFoodNet2012/index.html (accessed 18 April 2013).

Certified Organic Associations of British Columbia (COABC) (2009), 'British Columbia Certified Organic Production Operation Policies and Management Standards. Version 9. Book 2', available at http://certified organic.bc.ca/standards/docs/Book_2_V9.pdf (accessed 24 October 2013).

Chiappe, Maria B. and Cornelia Butler Flora (1998), 'Gendered Elements of the Alternative Agriculture Paradigm', *Rural Sociology*, **63**(3), 372–93.

Clapp, Jennifer and Doris Fuchs (eds) (2009), *Corporate Power in Global Agrifood Governance*, Cambridge: MIT Press.

Clark, Lisa F., Camille D. Ryan and William A. Kerr (2014), 'Direct Democracy, State Governments and the Re-energized GMO Debate: Implications of California's Proposition 37', *AgBioForum*, **16**(3), 177–86.

Clark, Lisa F. (2007), 'Business As Usual? Corporatization and the Changing Role of Social Reproduction in the Organic Agro-Food Sector', *Studies in Political Economy*, **80**, 55–74.

Clarke, Adrienne, et al. (2001), *Living Organic: Easy Steps to An Organic Lifestyle*, Napersville: Sourcebooks, Inc.

Clarkson, Stephen (2004), 'Global Governance and the Semi-Peripheral State: The WTO and NAFTA as Canada's External Constitution', in S. Clarkson and M.G. Cohen (eds), *Governing Under Stress*, Toronto: Fernwood Press, pp. 198–255.

Clunies-Ross, Tracey (1990), 'Organic Food: Swimming Against the Tide', in Terry Marsden and Jo Little (eds), *Political, Social and Economic Perspectives on the International Food System*, Brookfield: Avebury Books, pp. 200–14.

Codex Alimentarius Commission (Codex) (2001), 'Codex Alimentarius: Organically Produced Foods', Rome: FAO/WHO Codex Alimentarius Commission, available at www.codexalimentarius.net (accessed 8 November 2013).

Codex Alimentarius Commission (Codex) (1999), *Report of the Twenty-Seventh Session of the Codex Committee on Food Labelling*, Ottawa, Canada, 27–30 April 1999. ALINORM 99/22A. May 1999. Rome.

Cohen, Marjorie Griffin (2007), 'Collective Economic Rights and International Trade Agreements: In the Vacuum of Post-National Capital Control', in Susan Boyd and Margot Young (eds), *Poverty: Rights, Social Citizenship and Governance*, Vancouver: UBC Press.

Cohn, Theodore H. (2003), *Global Political Economy: Theory and Practice*, Toronto: Longman Publishers.

Cohn, Theodore H. (2002), *Governing Global Trade: International Institutions in Conflict and Convergence*, Aldershot: Ashgate.

Coleman, William D. and Austina Reed (2007), 'Legalisation Transnationalism and the Global Organic Movement', in Christian Bruetsch and Dirk Lehmkuhl (eds.), *Law and Legalisation in Transnational Relations*, London: Routledge, pp. 101–20.

Coleman, William D., et al. (2004), *Agriculture in the New Global Economy*, Northampton, MA, USA and Cheltenham, UK: Edward Elgar.

Commins, Ken (2003a), 'Comparison of the Requirements of the IFOAM Accreditation Criteria and the Requirements of ISO/IEC Guide 65, IFOAM', available at http://r0.unctad.org/trade_env/test1/projects/itf/IAC-ISO65comparison.pdf (accessed 19 March 2013).

Commins, Ken (2003b), 'IFOAM Normative Documents', in Christina Westermayer and Berwand Greier (eds), *The Organic Guarantee System: The Need and Strategy for Harmonisation and Equivalence*, Bonn: IFOAM, pp. 78–81.

Commission for Environmental Cooperation (CEC) (2000), 'Final Factual Record for Submission SEM-97-001, BC Aboriginal Fisheries Commission et al.', available at http://www.cec.org/Storage/68/6220_BC-Hydr-Fact-record_en.pdf (accessed 4 September 2013).

Conford, Philip (2001), *The Origins of the Organic Movement*, Edinburgh: Floris Books.

Convention on Biological Diversity (CBD) (2012), 'Frequently Asked Questions (FAQs) on the Cartagena Protocol', available at https://bch.cbd.int/protocol/cpb_faq.shtml#faq1 (accessed 3 October 2012).

Cooper, Rachelle (2006), 'Study Explores History of Health Food Stores, *At Guelph*, **50**(12), June 14', available at www.uoguelph.ca/atguelph/06-06-14/featureshealth.shtml (accessed May 2012).

Council of Canadians (2005), 'GE Free Canada Campaign–Council of Canadians', Public Forum. 2 June 2005. Maritime Labour Centre, Vancouver.

Council of the European Communities (1993), 'Council Regulation EEC No. 2092/91: On Organic Production of Agricultural Products', available at http://eurlex.europa.eu/LexUriServ/LexUriServ.do?uri=OJ:L:1991:198:0001:0015:EN:PDF (accessed 4 July 2013).

Cox, Robert (1999), 'Civil Society at the Turn of the Millennium: Prospects for an Alternative World Order', *Review of International Studies*, **25**(1), 3–28.

Coy, Patrick and Timothy Hedeen (2005), 'A Stage Model of Social Movement Co-optation: Community Mediation in the United States', *The Sociological Quarterly*, **46**, 405–35.

Cuddeford, Vijay (2004), 'When Organics Go Mainstream, *Cyber-Help for Organic Farmers*', available at http://www.certifiedorganic.bc.ca/rcbtoa/services/organics-mainstream.html (accessed 12 February 2014).

Dabbert, Stephan (2003), 'Organic Agriculture and Sustainability: Environmental Aspects', *Organic Agriculture: Sustainability, Markets and Policies*, OECD, Paris: CABI Publishing, pp. 51–64.

Daily Democrat (2005), 'Farm Workers Not Being Aided by Organic Growers', www.dailydemocrat.com/cda/article (accessed 29 April 2005).

Dahlberg, Kenneth (1979), *Beyond the Green Revolution: The Ecology and Politics of Global Agricultural Development*, New York: Plenum Press.

Dalton, Russel J. (1990), 'The Challenge of New Movements', in Russell J. Dalton and Manfred Kuechler (eds), *Challenging the Political Order: New Social and Political Movements in Western Democracies,* Cambridge: Polity Press, pp. 3–20.

Dalton, Russel J. and Manfred Kuechler (eds) (1990), *Challenging the Political Order: New Social and Political Movements in Western Democracies*, Cambridge: Polity Press.

Daniels, Stevie O. (1990), 'Organic Foods Act', *Organic Gardening*, **37**(5), 7.

Daugbjerg, Carsten (2012), 'The World Trade Organization and Organic Food Trade: Potential for Restricting Protectionism?' *Organic Agriculture*, **2**, 55–66.

DeLind, Laura B. (2000), 'Transforming Organic Agriculture into Industrial Organic Products: Reconsidering National Organic Standards', *Human Organization,* **59**(2), 198–209.

della Porta, Donatella and Sidney Tarrow (2005), 'Transnational Processes and Social Activism: An Introduction', in Donatella della Porta and Sidney Tarrow (eds), *Transnational Protest and Global Activism*, Toronto: Rowman & Littlefield Publishers, pp. 1–20.

Dicken, Peter (1999), *Global Shift: Transforming the World Economy*, London: Paul Chapman Publishing Ltd.

Desmarais, Annette Aurelie, Nettie Weibe and Hannah Wittman (eds) (2011), *Food Sovereignty in Canada: Creating Just and Sustainable Food Systems*, Toronto: Fernwood Publishing.

Dimitri, Carolyn and Catherine Greene (2002), 'Recent Growth Patterns in the U.S. Organic Foods Market', *Economic Research Service, USDA*, September.

Dimitri, Carolyn and Nessa J. Richman (2000), *Organic Food Markets in Transition*, Greenbelt, MD: Henry Wallace Center For Agricultural & Environment Policy.

Doherty, Paddy (2003), 'The Canadian Organic System as of March 2003', available at http://organic.usask.ca/ORC%20Updates/ORC% 20Update%20April03.pdf (accessed 10 February 2014).

Doyran, Selma H. (2003), 'Codex Guidelines on the Production Processing, Labelling and Marketing of Organically Produced Foods', in Christina Westermayer and Berwand Greier (eds), *The Organic Guarantee System: The Need and Strategy for Harmonisation and Equivalence*, Bonn: IFOAM, pp. 30–6.

Draffin, George (2006), 'The Incorporation of the Organic Food Industry', available at http://www.endgame.org/organics.html (accessed 4 August 2009).

Earthbound Farm (2012), 'The Earthbound Story', available at http:// www.ebfarm.com/story (accessed 14 September 2012).

The Earthcare Group (2010), 'The Earthcare Group', available at http:// www.econet.sk.ca/sk_enviro_champions/earthcare.html (accessed 11 July 2010).

Egri, Carolyn (1994), 'Working with Nature: Organic Farming and Other Forms of Resistance to Industrialized Agriculture', in John M. Jermier et al. (eds), *Resistance and Power in Organizations*, New York: Routledge, pp. 128–66.

Environment Canada (2012), 'National Air Pollutant Emissions', available at http://www.ec.gc.ca/indicateurs-indicators/default.asp?lang=en &n=E79F4C12-1 (accessed 4 November 2012).

Environmental Protection Agency (US) (2000), '2000 National Water Quality Inventory', available at http://water.epa.gov/lawsregs/guidance/ cwa/305b/upload/2003_02_28_30 5b_2000report_chp3.pdf (accessed 20 June 2014).

Euromonitor International (2013), 'Organic Packaged Food in the US', Passport GMID.

European Commission, Agriculture and Rural Development (2012a), 'U.S. – European Union Organic Equivalence Agreement Frequently Asked Questions and Answers', available at www.organic-farming. europa.eu (accessed 14 September 2012).

European Commission: Agriculture and Rural Development (2012b), 'Organic Farming: Specifics of EU Organic Legislation', available at http://ec.europa.eu/agriculture/organic/eu-policy/legislation_en#regulation (accessed 14 September 2012).

European Commission: Agriculture and Rural Development (2012c), 'Agreement between EU and Canada on Equivalency in Organic Products', available at http://ec.europa.eu/agriculture/newsroom/45_ en.htm (accessed 14 September 2012).

Esch, Mary (2013), 'Naturally Grown: An Alternative Label to Organic, Associated Press. August 17, 2013', available at http://news.yahoo.com/

naturally-grown-alternative-label-organic-160014399.html (accessed 22 August 2013).

FamilyFarmDefenders (2007), 'Welcome to Whole Foods The Walmart Of Organic', available at http://web.archive.org/web/20071011021114/ http://www.familyfarmdefenders.org/pmwiki.php/LocalFoodSystems/ WelcomeToWholeFoodsTheWalmartOfOrganic (accessed 16 January 2013).

FarmFolk/CityFolk (2013), 'Who We Are', available at http://www. ffcf.bc.ca/About_Us.html (accessed 21 May 2013).

Farmworker Justice, 'H-2a Guestworker Program. Advocacy Program', available at http://www.farmworkerjustice.org/node/56 (accessed 28 August 2013).

Finland, Ministry of Trade and Industry (1992), 'Notification to the Committee on Technical Barriers to Trade, TBT/Notification 92.44 (Feb. 25, 1992)', available at http://www.wto.org/gatt_docs/english/ sulpdf/91600392.pdf (accessed 19 July 2012).

Food and Agriculture Organization of the United Nations (FAO) (2010), 'The Principle Aims of Organic Agriculture and Processing', available at http://www.fao.org/ag/againfo/programmes/en/lead/toolbox/Tech/ OrganicA.htm (accessed 22 June 2013).

FAO (2005), 'FAO AGROSTAT Database', available at www.faostat.fao. org/ (accessed 8 September 2010).

FAO (2003), 'Chapter 11.3: Organic Agriculture, *World Agriculture: Towards 2015/2030: An FAO Perspective*,' available at www.fao.org/ documents/show_cdr.asp?url_file=/DOCREP/Y4252E?Y4252E00.htm (accessed 15 November 2010).

FoodSafety.gov (2015), 'Food Recalls', available at http://search.food safety.gov/search?q=organic&btnG.x=21&btnG.y=16&btnG=Search& sort=date%3AD%3AL%3Ad1&client=food-safety&entqr=0&oe=UTF-8&ie=UTF-8&ud=1&proxystylesheet=food-safety&output=xml_no_dt d&site=food-safety&lr=lang_en (accessed 15 January 2015).

Forbes, Linda C. and John M. Jermier (2002), 'The Institutionalization of Voluntary Organizational Greening and the Ideals of Environmentalism', in Andrew J. Hoffman and Marc J. Ventresca (eds), *Organizations, Policy and the Natural Environment, Institutional and Strategic Perspectives*, Palo Alto: Stanford University Press, pp. 194–213.

Forbes Magazine (2010), 'Personal Finance: America's Biggest Food Companies, 11/02/2010 @ 10:13 AM', available at http://www.forbes. com/sites/investopedia/2010/11/02/americas-biggest-food-companies/3/ (accessed 10 September 2012).

Freyfogle, E. (2001), *The New Agrarianism: Land, Culture and Community of Life*, Washington DC, Island Press.

Fridell, Gavin (2007), *Fair Trade Coffee: The Prospects and Pitfalls of Market-Driven Social Justice*, Toronto: University of Toronto Press.

Friedland, William H. (1994), 'The New Globalization: The Case of Fresh Produce', in Alessandro Bonanno et al. (eds), *From Columbus to ConAgra: The Globalization of Agriculture*, Lawrence: University Press of Kansas, pp. 210–31.

Friedmann, Harriet (2007), 'Scaling Up: Bringing Public Institutions and Food Service Corporations into the Project for a Local, Sustainable Food System in Ontario', *Agriculture and Human Values*, **24**, 389–98.

Friedmann, Harriet (2000), 'What on Earth is the Modern Food System? Foodgetting and Territory in the Modern Era and Beyond', *Journal of World Systems Research*, **11**(2), 480–515.

Friedmann, Harriet (1991), 'Changes in the International Division of Labour. Agri-Food Complexities and Export Agriculture', in William Friedland (ed.), *Towards a New Political Economy of Agriculture*, Boulder: Westview Press, pp. 65–93.

Friedmann, Harriett (1978), 'World Market, State, and Family Farm: Social Bases of Household Production in the Era of Wage Labor', *Comparative Studies in Society and History*, **20**(4), 545–86.

Friedmann, Harriet and Philip McMichael (1989), 'Agriculture and the State System: The Rise and Decline of National Agricultures, 1870 to the present', *Sociologia Ruralis*, **2**, 93–117.

Fromartz, Samuel (2006), *Organic Inc.: Natural Foods and How they Grew*, New York: Harcourt Press.

Gendron, Corinne et al. (2006), 'The Institutionalization of Fair Trade: More than a Degraded Form of Social Action, Les cahiers de Chaire-collection recherché, No. 12. Ecole des sciences de la gestion, Universite du Quebec a Montreal', available at http://www.crsdd.uqam.ca/pdf/pdfCahiersRecherche/2006/12-2006.pdf (accessed 5 May 2011).

Gereffi, Geoffrey and Miguel Korezeniewicz (1994), 'Introduction: Global Commodity Chains', in G. Gereffi and M. Korezeniewicz (eds), *Commodity Chains and Global Capitalism*, Westport, CT: Praeger, pp. 1–14.

Getz, C., S. Brown and A. Shreck (2008), 'Class Politics and Agricultural Exceptionalism in California's Organic Movement', *Politics and Society*, **36**, 478–507.

Giangrande, Carole (1985), *Down to Earth: The Crisis in Canadian Farming*, Toronto: Anansi Press Ltd.

Gillon, Sean (2011a), 'Appropriationism', *Green Food: An A-to-Z Guide*, London: SAGE Publications.

Gillon, Sean (2011b), 'Substitutionalism', *Green Food: An A-to-Z Guide,* London: SAGE Publications.

Gilman, Robert (1990), 'Sustainability: The State Of The Movement', *Context Institute,* Spring Issue, 10.

Global Organic Market Access (GOMA) (2012), 'Harmonization and Equivalence: The Newsletter of GOMA', available at http://www.goma-organic.org/ (accessed 13 October 2012).

Glover, Paul (2005), 'What We Need To Know About the Corporate Takeover of the "Organic" Food Market', available at http://freepage.twoday.net/stories/1827407/ (accessed 3 August 2009).

GMO Compass (2012), 'Crops: Soybeans', available at http://www.gmo-compass.org/eng/grocery_shopping/crops/19.genetically_modified_soy bean.html (accessed 3 October 2012).

Gold, Mary (2005), 'Organic Agriculture Products: Marketing and Trade Resources, *Agricultural Research Service, USDA*', available at http:// www.nal.usda.gov/afsic/AFSIC_pubs/OAP/srb0301a.htm (accessed 9 October 2010).

Goldman, M.C. (1970), 'Southern California: Food Shopper's Paradise', *Organic Gardening and Farming,* November, 38–45.

Goldstein, Jermone (1976), 'Organic Force', in Richard Merrill (ed.), *Radical Agriculture,* New York: New York University Press, pp. 212–23.

Goodman, David (2000), 'Organic and Conventional Agriculture: Materializing Discourse and Agro-ecological Managerialism', *Agriculture and Human Values,* **17**, 215–19.

Goodman, David and E. Melanie Dupuis (2002), 'Knowing Food and Growing Food: Beyond the Production-Consumption Debate in the Sociology of Agriculture', *Sociologia Ruralis,* **42**(1), 5–22.

Goodman, David, Bernardo Sorj and John Wilkinson (1987), *From Farming to Biotechnology: A Theory of Agro-Industrial Development,* Oxford: Blackwell Publications.

Granovetter, Mark (1985), 'Economic Action and Social Structure: The Problem of Embeddedness', *American Journal of Sociology,* **91**(3), 481–510.

Green, Catherine and Amy Kremen (2003), 'U.S. Organic Farming in 2000–2001: Adoption of Certified Systems', *Agriculture Information Bulletin No. AIB780, April,* available at www.ers.usda.gov/Publications/ AIB780 (accessed 2 June 2013).

Greene, Wade (1971), 'Guru of the Organic Food Cult', *New York Times,* 6 June, SM30. ProQuest Historical Newspapers. *NYT,* (1851–2003).

Guidry, John A. (2000), 'Globalization and Social Movements', in John A. Guidry, M. Kennedy and M. Zald (eds), *Globalizations and Social*

Movements: Culture, Power and the Transnational Public Sphere, Ann Arbor: The University of Michigan Press.

Guthman, Julie (2004), *Agrarian Dreams: The Paradox of Organic Farming in California*, Berkeley: University of California Press.

Guthman, Julie (2003), 'Fast Food/Organic Food: Reflexive Tastes and the Making of "Yuppie Chow"', *Social and Cultural Geography*, **4**(1), 45–58.

Haas, Ernst B. (1990), *When Knowledge is Power: Three Models of Change in International Organization*, Berkeley: University of California Press.

The Hain-Celestial Group Inc. (2013), 'Annual Report', available at www.hain-celestial.com (accessed 8 January 2015).

The Hain-Celestial Group Inc. (2012), 'Annual Report', available at www.hain-celestial.com (accessed 8 January 2015).

The Hain-Celestial Group Inc. (2005), 'Annual Report', available at www.hain-celestial.com (accessed 13 October 2006).

Hall, Alan and Veronika Mogyorody (2001), 'Organic Farmers in Ontario: An Examination of the Conventionalization Argument', *Sociologica Ruralis* **41**(4), 399–422.

Hall, Stuart (1991), 'Brave New World', *Socialist Review*, **21**(1), 57–64.

Hall, Stuart B. (2014), 'Chapter 22: Considerations for Enabling the Ecological Redesign of Organic and Conventional Agriculture: A Social Ecology and Psychosocial Perspective', in Stephane Bellon and Servane Penvern (eds), *Organic Farming, Prototype for Sustainable Agricultures*, New York: Springer, pp. 401–22.

Hallam, David (2003), 'The Organic Market in OECD countries: Past Growth, Current Status and Future Potential', *Organic Agriculture: Sustainability, Markets and Policies*, OECD, Paris: CABI Publishing, pp. 179–86.

Hancock, Michelle (2006), 'Canadian Organic Regulations, *Alive Magazine*', available at http://www.alive.com/articles/view/20499/canadian_organic_regulations (accessed 15 October 2012).

Hansenclever, Andreas et al. (1997), *Theories of International Regimes*, Cambridge: Cambridge University Press.

Harris, Mark T. (2006), 'Welcome to "Whole-Mart": Rotten Apples in the Social Responsibility Industry', *Dissent*, Winter, 61–6.

Harrison, John B. (1993), *Growing Food Organically*, Vancouver: Waterwheel Press.

Harter, Walter (1973), *Organic Gardening for the City Dweller*, New York: Warner Press.

Harwood, Richard (1984), 'Organic Farming Research at the Rodale Center', in D.F. Bezdicek et al. (eds), *Organic Farming: Current*

Technology and Its Role in Sustainable Agriculture, Madison: Soil Science Association of America, pp. 1–17.

Haumann, Barbara Fitch (2010), 'North America', in Helga Willer and Lukas Kilcher (eds.), *The World of Organic Agriculture: Statistics and Emerging Trends 2010*, Research Institute of Organic Agriculture (FiBL), Frick, and International Federation of Organic Agriculture Movements (IFOAM), Bonn, pp. 184–92.

Haumann, Barbara Fitch (2014), 'Another Milestone Year for the US Organic Industry', Helga Willer and Julia Lernoud (eds), *The World of Organic Agriculture – Statistics and Emerging Trends 2014*, Research Institute of Organic Agriculture (FiBL), Frick, and International Federation of Organic Agriculture Movements (IFOAM), Bonn: 241–7.

Hawaleshka, Danylo, Brian Bethune and Sue Ferguson (2004), 'Tainted Food', *Maclean's*, 26 January, pp. 22–30.

Hechter, Michael (2004), 'From Class to Culture', *American Journal of Sociology*, **110**, 400–45.

Heffernan, William H. and Mary Hendrickson (2002), 'Multi-National Concentrated Food Processing and Marketing Systems and The Farm Crisis', presented at the Annual Meeting of the American Association for the Advancement of Science Symposium, Boston: 14–19 February.

Heffernan, William H. et al. (1999), 'Consolidation in the Food and Agriculture System, Report to the National Farmers Union', available at http://www.foodcircles.missouri.edu/whstudy.pdf (accessed 20 June 2014).

Heffernan, William H. and D.H. Constance (1994), 'Transnational Corporations and the Globalization of the Food System', in Alessandro Bonanno et al. (eds), *From Columbus to ConAgra: The Globalization of Agriculture*, Lawrence: University Press of Kansas, pp. 29–51.

Heffernan, William D. (1998), 'Agriculture and Monopoly Capital', *Monthly Review*, **50**(3), 46–59.

Henderson, Elizabeth (1998), 'Rebuilding Local Food Systems From the Grassroots Up', *Monthly Review*, **50**(3), 112–24.

Hendrickson, Mary and William Heffernan (2007), 'Concentration of Agricultural Markets April 2007', available at http://www.foodcircles.missouri.edu/07contable.pdf (accessed 26 April 2010).

Henry, M. (2001), 'Sow Few, So Trouble', *Green Futures,* **31**, 40–42.

Hewitt, Jean (1970), 'Organic Food Fanciers Go to Great Lengths for the Real Thing', *New York Times* 7 September, 23.

Hill, Stuart and Rod J. MacRae (1992), 'Chapter 1', *Organic Farming in Canada*, Montreal: McGill University, Ecological Agricultural Projects publications.

Hobbs, Jill E. (2001a), 'Developing Supply Chains for Nutraceudicals and Functional Foods: Opportunities and Challenges', paper presented at INAF/CREA, 23 November, Laval University, available at http://www4.agr.gc.ca/resources/prod/doc/misb/fb-ba/nutra/ffn-afn_e.pdf (accessed 15 October 2011).

Hobbs, Jill E. (2001b), 'Labelling and Consumer Issues in International Trade', in Hans J. Michelmann et al. (eds), *Globalization and Agricultural Trade*, Boulder: Lynne Rienner Publishers, pp. 269–85.

Hodess, Robin (2001), 'The Contested Competence of NGOs and Business in Public Life', in Daniel Drache (ed.), *The Market or the Public Domain: Global Governance & the Asymmetry of Power*, New York: Routledge, pp. 129–47.

Hoekman, Bernard and Michael Kostecki (2001), *The Political Economy of the World Trading System*, New York: Oxford University Press.

Holmes, Matthew and Ann Macey (2014), 'Organic Agriculture in Canada', in Helga Willer and Julia Lernoud (eds), *The World of Organic Agriculture – Statistics and Emerging Trends 2014*, Research Institute of Organic Agriculture (FiBL), Frick, and International Federation of Organic Agriculture Movements (IFOAM), Bonn, pp. 247–50.

Hoodes, Liana et al. (2010), 'National Organic Action Plan: From the Margins to the Mainstream – Advancing Organic Agriculture in the U.S., Rural Advancement Foundation International – USA', available at http://rafiusa.org/docs/noap.pdf (accessed 19 January 2015).

Hoppe, Robert A. and David E. Banker (2006), 'Structure and Finances of US Farms: 2005 Family Farm Report, USDA, ERS (May)', available at http://151.121.68.30/publications/EIB12/EIB12_reportsummary.pdf (accessed 6 July 2013).

Hoppe, Robert A. et al., (2004), 'Differences in Canadian and U.S. Farm Structure: What the Canadian Farm Typology Shows', *Current Agriculture, Food & Resource Issues* **4**, 83–94.

Howard, Sir Albert (1947), *The Soil and Health: A Study of Organic Agriculture,* New York: Devin-Adair Company.

Howard, Sir Albert (1946), *War in the Soil*, Emmaus: Organic Gardening.

Howard, Sir Albert (1943), *An Agricultural Testament*, Oxford: Oxford University Press.

Howard, Patricia (2000), 'Genetic Modification of Foods and Seeds: Is It Inherently Dangerous?', *Canadian Dimension*, **34**(4).

Howard, Philip H. (2014), 'Organic Processing Industry Structure', available at www.msu.edu/~howardp/organicindustry.html (accessed 26 January 2014).

Howard, Philip H. (2013), 'Who Owns What? Cyber-Help for Organic Farmers', available at http://certifiedorganic.bc.ca/rcbtoa/services/corporate-ownership.html (accessed 12 February 2014).

Howard, Philip H. (2009a), 'Consolidation in the North American Organic Food Processing Sector, 1997 to 2007', *International Journal of Agriculture and Food,* **16**(1), 13–30.

Howard, Philip H. (2009b), 'Visualizing Food System Concentration and Consolidation', *Southern Rural Sociology*, **24**(2), 87–110.

Howard, Philip H. (2006), 'Consolidation in Food and Agriculture: Implications for Farmers & Consumers', *The Natural Farmers*, Spring, 17–20.

Human Resources and Skills Development Canada (HRSDC) (2010), 'Temporary Foreign Worker Program: Labour Market Opinion (LMO) Statistics, Annual Statistics 2006–2009 (Table 10)', available at http://www.hrsdc.gc.ca/eng/workplaceskills/foreign_workers/stats/annual/table 10a.shtml (accessed 5 September 2012).

Ikerd, John (1999), 'Organic Agriculture Faces the Specialization of Production Systems: Specialized Systems and the Economical Stakes', paper presented at Jack Cartier Centre, Lyon, France, 6–9 December, available at www.ssu.missouri.edu/faculty/jikerd/papers/FRANCE.html (accessed 23 September 2012).

Imhoff, Daniel (1996), 'Community Supported Agriculture: Farming with a Face on It', in Jerry Mander and Edward Goldsmith (eds), *The Case Against the Global Economy and For a Turn Towards the Local*, San Francisco: Sierra Club Books, pp. 425–33.

Ingelhart, Ronald (1990), 'Values, Ideology, and Cognitive Mobilization in New Social Movements', Russell J. Dalton and Manfred Kuechler (eds), *Challenging the Political Order: New Social Movements in Western Democracies*, Cambridge: Polity Press, pp. 43–66.

Ingelhart, Ronald and David Appel (1989), 'The Rise of Post-Materialist Values and Changing Religious Orientations, Gender Roles and Sexual Norms', *International Journal of Public Opinion Research,* **1**(1), 45–75.

Institute of Food Technologists (IFT) (1974), 'Organic Foods: A Scientific Status Summary by the Institute of Food Technologists', Expert Panel on Food Safety and Nutrition & Committee on Public Information', *Food Technology*, January.

International Federation of Organic Agricultural Movements (IFOAM) (2014), 'International Harmonization', available at http://www.ifoam.bio/fr/value-chain/harmonization-and-equivalence (accessed 12 July 2014).

IFOAM (2009), 'Definition of Organic Agriculture', available at https://web.archive.org/web/20081028104343/http://www.ifoam.org/growing_organic/definitions/doa/index.html (accessed 30 October 2012).

IFOAM (2007), 'The New EU Regulation for Organic Food and Farming: (EC) No 834/2007: Background, Assessment, Interpretation', available at https://www.fh-muenster.de/fb8/downloads/strassner/vero effentlichungen/2009_CM_MS_Eds_68s.pdf (accessed 6 September 2013).

IFOAM (2005a), 'IFOAM and the Codex Alimentarius Commission', available at www.ifoam.org/about_ifoam/status/codex.html (accessed 3 November 2009).

IFOAM (2005b), 'Recommendations for Inspection of Social Standards', available at http://www.naturland.de/fileadmin/MDB/documents/Publication/English/Social-Standards_Tools_and_Methodologies_of_Inspection.pdf (accessed 12 July 2013).

IFOAM (2005c), 'SASA: Social Accountability in Sustainable Agriculture', available at https://web.archive.org/web/20060903170442/http://www.ifoam.org/organic_facts/justice/sasa.html (accessed 12 July 2014).

IFOAM (2004), 'Basic Labour Requirements for Organic Traders and Suppliers, Options for Code of Conduct for IFOAM Traders', available at http://www.novotrade.nl/talk/paper/talk-01.html (accessed 15 November 2014).

IFOAM (2002a), 'A Social Agenda for Organic Agriculture?', available at www.ifoam/organic_facts/justice/pdfs/Social_Agenda_for_Organic_Agriculture.pdf (accessed 12 July 2007).

IFOAM (2002b), 'Sustainability and Organic Agriculture, Johannesburg: Position Paper, World Summit on Sustainable Development', available at http://biodiversityeconomics.org/pdf/020831.PDE (accessed 15 November 2007).

IFOAM (1999), 'IFOAM's position for the WTO Seattle Meeting December 1999', Bonn: IFOAM.

International Organization for Standardization (ISO) (2012), 'ISO/IEC 17065:2012. Conformity Assessment – Requirements for Bodies Certifying Products, Processes and Services', available at http://www.iso.org/iso/home/store/catalogue_ics/catalogue_detail_ics.htm?csnumber=46 568 (accessed 30 April 2013).

ISO (2006), 'Why Standards Matter, ISO and World Trade, Overview of the ISO', available at http://www.iso.org/iso/en/aboutiso/introduction/index.html (accessed 21 July 2007).

Jackson, Laura L. (1998), 'Agricultural Industrialization and the Loss of Biodiversity', in Lakshman Guruswamy and J.A. McNeely (eds), *Protection of Global Biodiversity: Converging Strategies*, Durham: Duke University Press, pp. 66–77.

Jacob, Jeffrey (1997), *New Pioneers: The Back to Land Movement and the Search for a Sustainable Future*, University Park, Penn: Pennsylvania State University Press.

Jacobsen, Birthe Thode (2002), 'Organic Farming and Certification, International Trade Centre: UNCTAD/WTO', available at http://www.hubrural.org/IMG/pdf/organic-farming-+certification.pdf (accessed 24 June 2008).

Jessop, Bob (1993), 'Towards a Schumpeterian Workfare State? Preliminary Remarks on Post-Fordist Political Economy', *Studies in Political Economy*, **40**, 7–39.

Johnson, Josee and Michelle Szabo (2011), 'Reflexivity and the Whole Foods Market Consumer: The Lived Experience of Shopping for Change', *Agriculture and Human Values*, **28**(3), 303–19.

Josling T, D. Roberts and D. Orden (2004), *Food Regulation and Trade: Toward Safe and Open Global System*, Washington DC: Institute for International Economics.

Josling, Tim (2001), 'Regional Trade Agreements and Agriculture: A Post-Seattle Assessment', in Hans J. Michelmann et al. (eds), *Globalization and Agricultural Trade*, Boulder: Lynne Rienner Publishers, pp. 171–95.

Justica (2013), 'Canada's Seasonal Agricultural Worker Programme', available at http://www.justicia4migrantworkers.org/bc/pdf/sawp.pdf (accessed 23 July 2013).

Kabel, Marcus (2006), 'Wal-Mart's Organics Could Shake Up Retail', available at http://www.myplainview.com/article_7f0d6c5f-cff6-573a-b7d2-2b3e97e9e8c3.html (accessed 24 March 2013).

Kaplinsky, Raphael (2000), 'Globalisation and Unequalisation: What Can Be Learned from Value Chain Analysis?', *The Journal of Development Studies*, **37**(2), 117–46.

Kaplinsky, Raphael and Mike Morris (2000), *A Handbook for Value Chain Research*, available at http://www.prism.uct.ac.za/Papers/VchNov01.pdf (accessed 17 June 2012).

Keck, Margaret and Kathryn Sikkink (2000), 'Historical Precursors to Modern Transnational Social Movements and Networks', *Globalizations and Social Movements: Culture, Power and the Transnational Public Sphere*, Ann Arbor: University of Michigan Press, pp. 35–53.

Keck, Margaret and Kathryn Sikkink (1998), *Activism Beyond Borders: Advocacy Networks in International Politics*, Ithaca: Cornwell University Press.

Keene, Paul (1988), *Fear Not to Sow*, Chester: The Globe Pequot Press.

Keith, Barbara (2003), 'More than Just Farming: Employment in Agriculture and Agri-Food in Rural and Urban Canada, *Rural and Small Town Canada Analysis Bulletin,* **4**(8) Statistics Canada', available at http://publications.gc.ca/collections/Collection/Statcan/21-006-X/21-006-XIE2002008.pdf (accessed 11 October 2013).

Kerton, Sarah and A. John Sinclair (2010), 'Buying Local Organic Food: A Pathway to Transformative Learning', *Agriculture and Human Values*, **27**, 401–13.

Kerr, William A. (2001), 'The World Trade Organization and the Environment', in Hans J. Michelmann et al. (eds), *Globalization and Agricultural Trade*, Boulder: Lynne Rienner Publishers, pp. 53–65.

Kilcher, Lukas et al. (2004), 'Standards and Regulations', in Helga Willer and Lukas Kilcher (eds), *The World of Organic Agriculture: Statistics and Emerging Trends 2004*, Research Institute of Organic Agriculture (FiBL), Frick, and International Federation of Organic Agriculture Movements (IFOAM), Bonn, pp. 27–43.

Kinchy, A. J. (2012), *Seeds, Science, and Struggle: the Global Politics of Transgenic Crops*, Boston: MIT Press.

Kirschenmann, Fred (2004), 'A Brief History of Sustainable Agriculture', *The Networker,* **9**(2), available at http://www.sehn.org/Volume_9-2.html #a2 (accessed 14 December 2014).

Klintman, Mikael and Magnus Bostrom (2013), 'Political Consumerism and the Transition Towards a More Sustainable Food Regime: Looking Behind and Beyond the Organic Shelf', in Gert Spaargaren, Peter Oosterveer and Anne Loeber (eds), *Food Practices in Transition: Changing Food Consumption, Retail and Production in the Age of Reflexive Modernity*, Routledge, pp. 107–52.

Klonsky, Karen and Kurt Richter (2011), 'Statistical Review of California's Organic Agriculture, 2005–2009', available at http://aic.ucdavis.edu/publications/Statistical_Review_05-09.pdf (accessed 3 March 2012).

Klonsky, Karen and Kurt Richter (2007), 'Statistical Review of California's Organic Agriculture, 2000–2005', available at http://aic.ucdavis.edu/publications/Statistical_Review_00-05.pdf (accessed 12 July 2011).

Klonsky, Karen (2000), 'Forces Impacting the Production of Organic Foods', *Agriculture and Human Values*, **17**, 233–43.

Kloppenburg, Jack (ed.) (1988), *Seeds and Sovereignty: The Use and Control of Plant Genetic Resources*, Durham: Duke University Press.

Kneen, Brewster (1989), *From Land to Mouth: Understanding the Food System*, Toronto: NC Press.

Knutson, Ronald et al. (2007), *Agricultural Food Policy*, Columbus: Pearson Prentice Hall.

Kortbech-Olesen, Rudy (2004), *The Canadian Market for Organic Foods and Beverages*, International Trade Centre Paris: UNCTAD/WTO.

Korten, David (1995), *When Corporations Rule the World*, West Haven: Kumarian Press.

Kratochwil, Fredrick and John G. Ruggie (1986), 'International Organizations: The State of the Art on the Art of the State', *International Organization*, **40**(4), 753–76.

Kuepper, George and Lance Gegner (2004), 'Organic Crop Production Overview, ATTRA: National Sustainable Agriculture Information Service', available at http://attra.ncat.org/attra-pub/organiccrop.html (accessed 18 November 2010).

Lacy, Michael G. (1982), 'A Model of Cooptation Applied to the Political Relations of the United States and American Indians,' *Social Science Journal*, **19**, 24–36.

Lampkin, Nicholas H. and Susanne Padel (eds) (1994), *The Economics of Organic Farming: An International Perspective*, Wallingford: CAB International.

Lampkin, Nicholas H. (1994), 'Researching Organic Farming Systems', in Nicholas Lampkin and Susanne Padel (eds), *The Economics of Organic Farming: An International Perspective*, Wallingford: CAB International, pp. 27–43.

Lang, Tim et al. (2006), 'The Food Industry, Diet, Physical Activity and Health: A Review of Reported Commitments and Practice of 25 of the World's Largest Food Companies, April, Centre for Food Policy, London', available at http://image.guardian.co.uk/sys-files/Guardian/documents/2006/04/10/foodreportbig25.pdf (accessed 25 April 2013).

Lang, Tim and Michael Heasman (2005), *Food Wars: The Global Battle for Mouths, Minds and Markets*, London: Earthscan.

Lang, Tim (2004), 'Food Industrialization and Food Power: Implications for Food Governance', Simon Maxwell and Rachel Slater (eds), *Food Policy Old and New*, Oxford: Blackwell Publishing, pp. 21–31.

Lang, Tim (2003), 'Battle of the Food Chain', *The Guardian: Food – Why We Eat This Way*, 17 May, pp. 18–19.

Lappe, Frances Moore (1971), *Diet for a Small Planet*, New York: Friends of the Earth/Ballentine Books.

Lazarus, Eve (2010), 'Canadian Grocer, "Even in Hard Times, Organic is Hot"', available at http://www.canadiangrocer.com/categories/natural-selection-249 (accessed 23 October, 2012).

Lernoud, Julia, Helga Willer and Bernard Schlatter (2014), 'North America: Current Statistics', in Helga Willer and Julia Lernoud (eds), *The World of Organic Agriculture – Statistics and Emerging Trends 2014*, Research Institute of Organic Agriculture (FiBL), Frick, and IFOAM, Bonn, pp. 251–4.

Levenstein, Harvey (1993), *The Paradox of Plenty: A Social History of Eating in Modern America*, New York: Oxford University Press.

Lilliston, Ben and Ronnie Cummins (1998), 'Organic Versus "Organic": The Corruption of a Label', *The Ecologist* July 1998, available at https://www.organicconsumers.org/old_articles/Organic/orgvsorg.htm (accessed 24 May 2014)

Lindblom, Charles (1977), *Politics and Markets: The World's Political Economic Systems*, New York: Basic Books.

Lipschutz, Ronny D. (1992), 'Reconstructing World Politics: The Emergence of Global Civil Society', *Millennium*, **21**(3), 389–420.

Lipson, Elaine (2004), 'Food, Farming ... Feminism? Why Going Organic Makes Good Sense, (Summer Feature) *Ms. Magazine*,' available at www.msmagazine.com/summer2004/organicfarming.asp (accessed 10 September 2014).

Lockeretz, William (2007), 'What Explains the Rise of Organic Farming?' in William Lockeretz (ed.), *Organic Farming: An International History*, Wallingford: CABI, pp. 1–8.

Logsdon, Gene (1993), 'National Organic Standards: Kiss of Death? *The New Farm Classic* March (reprinted in 2005) The Rodale Institute', available at http://www.newfarm.org/depts/nf_classics/0604/logsdon.shtml (accessed 6 April 2013).

Lohr, Luanne and Barry Krissoff (2000), 'Consumer Effects of Harmonizing International Standards for Trade in Organic Foods', in Barry Krissoff et al. (eds), *Global Food Trade and Consumer Demand for Quality*, New York: Kluwer Academic Publishers, pp. 209–28.

Lohr, Luanne (1998), 'Implications of Organic Certification for Market Structure and Trade', *American Journal of Agricultural Economics*, **80**(5), 1125–9.

Lynch, Derek (2009), 'Environmental Impacts of Organic Agriculture: A Canadian Perspective', *Canadian Journal of Plant Science*, **89**, 621–8.

Macey, Anne (2007), *Retail Sales of Certified Organic Food Products in Canada in 2006*, Truno NS: Organic Agriculture Centre of Canada.

Macey, Anne (2004), *'Certified Organic', The Status of the Canadian Organic Market in 2003*, Ottawa: Report to Agriculture and Agri-Food Canada.

MacRae, Rod et al. (2004), **'**Does the Adoption of Organic Food and Farming Systems Solve Multiple Policy Problems? A Review of the Existing Literature', available at http://www.organicagcentre.ca/DOCs/Paper_Benefits_Version2_rm.pdf (accessed 26 March 2012).

MacRae, Rod (1990), 'History of Sustainable Agriculture: Strategies for Overcoming the Barriers to the Transition to Sustainable Agriculture', PhD thesis, available at http://www.eap.mcgill.ca/AASA_1.htm (accessed 20 June 2013).

Mallet, Patrick (2003), 'Options for Accreditation: National and International Accreditation Systems', in Christina Westermayer and Berwand Greier (eds), *The Organic Guarantee System: The Need and Strategy for Harmonisation and Equivalence*, Bonn: IFOAM, pp. 85–92.

Mann, Susan Archer (1990), *Agrarian Capitalism in Theory and Practice*, Chapel Hill: University of South Carolina Press.

Mark, Jason (2006), 'Workers on Organic Farms are Treated as Poorly as Their Conventional Counterparts', Grist.org, 2 August 2006 11:00PM, available at http://grist.org/article/mark/ (accessed 5 August 2012).

Mark, Jason (2004), 'Big Business Follows the Green', *AlterNet* Aug, available at http://www.alternet.org/story/19645/big_business_follows_the_green (accessed 2 August 2011).

Marsden, Terry et al. (1996), 'Agricultural Geography and the Political Economy Approach: A Review', *Economic Geography,* **72**(4), 361–75.

Marshall, W.E. (1974), 'Health Food, Organic Foods, Natural Foods: What They Are and What Makes Them Attractive to Consumers?', *Food Technology*, **28**(2), 50–56.

Marter, Marilynn (1989), 'Organic Food: Cost Is High, Demand Varies', *The Inquirer,* 21 June, available at http://articles.philly.com/1989-06-21/food/26106892_1_organic-products-organic-foods-foods-and-meats (accessed 18 September 2011).

Martin, Andrew (2007), 'Whole Foods Makes Offer for Small Rival', *The New York Times* [business section] Feb. 22', available at www.nytimes.com (accessed 23 February 2012).

McAdam, Doug, John D. McCarthy and Mayer N. Zald (1996), 'Introduction: Opportunities, Mobilizing Structures, and Framing Processes – Toward a Systematic, Comparative Perspective on Social Movements', in Doug McAdam et al. (eds), *Comparative Perspectives on Social Movements: Political Opportunities, Mobilizing Structures and Cultural Framings*, Cambridge: Cambridge University Press, pp. 1–22.

McAdam, Doug (1983), 'The Decline of the Civil Rights Movement', in Jo Freeman (ed.), *Social Movements of the Sixties and Seventies*, New York: Longman.

McAdam, Doug (1982), *Political Process and the Development of Black Insurgency, 1930-1970*, Chicago: University of Chicago Press.

McBean, L.D. and E.W. Speckmann (1974), 'Food Faddism: A Challenge to Nutritionist and Dietitians', *American Journal of Clinical Nutrition*, **27**, 1071–8.

McBride, Stephen (2001), *Paradigm Shift: Globalization and the Canadian State*, Halifax: Fernwood Publishing.

McGrath, Mike (1991), 'The Bashin' of the Green (Or Kiss Me, I'm Organic)', *Organic Gardening*, April, pp. 5–6.

McLeod, D. (1976), 'Urban Rural Food Alliances: A Perspective on Recent Community Food Organizing', in Richard Merrill (ed.), *Radical Agriculture*, New York: Harper Colophon Books, pp. 188–211.

McMichael, Philip (2004), 'Introduction: Agri-food System Restructuring: Unity in Diversity', *Global Development and the Corporate Food Regime*, paper presented at the XI World Congress of Rural Sociology, Trondheim, 1–18.

McMichael, Philip (2003), 'Food Security and Social Reproduction: Issues and Contradictions', in Stephen Gill and Isabella Bakker (eds), *Power, Production and Social Reproduction: Human In/Security in the Global Political Economy*, New York: Palgrave-MacMillan, pp. 169–89.

McNeely, J.A. and S.J. Scherr (2001), *Common Ground, Common Future: How Ecoagriculture can Help Feed the World and Save Wild Biodiversity*, Washington, D.C: IUCN, Future Harvest.

Meadows, Donnella H. et al. (1972), *Limits to Growth,* New York: Universe Books.

Mearnes, Alison C. (1997), 'Making the Transition from Conventional to Sustainable Agriculture: Gender, Social Movement Participation, and Quality of Life on the Family Farm', *Rural Sociology*, **62**(1), 21–47.

Merrigan, Kathleen (2003), 'The Role of Government Standards and Market Facilitation', *Organic Agriculture: Sustainability, Markets and Policies*, Paris: OECD Publications, pp. 280–97.

Merrill, Richard (ed.) (1976), *Radical Agriculture*, New York: Harper Colophon Books.

Meyer, David S. and Sidney Tarrow (1998), 'A Movement Society: Contentious Politics for a New Century', in David S. Meyer and Sidney Tarrow (eds), *The Social Movement Society: Contentious Politics for a New Century*, Oxford: Rowman and Littlefield Publishers, pp. 1–28.

Michelsen, Johannes (2001a), 'Organic Farming in a Regulatory Perspective', *Sociologica Ruralis*, **41**(1), 65–84.

Michelsen, Johannes (2001b), 'Recent Development and Political Acceptance of Organic Farming in Europe', *Sociologica Ruralis*, **41**(1), 3–20.

Michelsen, Johannes and V. Soregaard (2002), 'Policy Instruments for Promoting Conversion to Organic Farming and Their Effects in Europe 1985–97', *Skriftserie,* Esbjerg: Department of Political Science and Public Management.

Miele, Mara and Jonathan Murdoch (2002), 'The Practical Aesthetics for Traditional Cuisines: Slow Food in Tuscany', *Sociological Ruralis*, **42**(4), 312–28.

Miller, James C. and Keith H. Coble (2005), 'Cheap Food Policy: Fact or Rhetoric?', paper presented at the American Agricultural Economics Association Annual Meeting, 24–27 July, 2005.

Millstone, Erik and Tim Lang (2003), *The Penguin Atlas of Food: Who Eats What and Why*, Harmondsworth: Penguin Books.

Mitchell, Don (1975), *The Politics of Food*, Toronto: Lorimer Press.

Mutersbaugh, Tad (2005) 'Fighting Standards with Standards: Harmonization, Rents and Social Accountability in Certified Agrofood Networks', *Environment and Planning*, **37**(11), 2033–51.

Mutersbaugh, Tad (2002), 'The Number is the Beast: A Political Economy of Organic-Coffee Certification and Producer Unionism', *Environment and Planning A*, **34**, 1165–84.

Murdoch, Jonathan and Mara Miele (1999), 'Back to Nature: Changing "Worlds of Production" in the Food Sector', *Sociologia Ruralis*, **39**(4), 465–83.

Myers, Robin (1976), 'The National Sharecroppers Fund and the Farm Co-op Movement in the South', in Richard Merrill (ed.), *Radical Agriculture*, New York: Harper Colophon Books, pp. 129–42.

National Farmers Union (NFU) (2013), 'National Farmers Union Policy on Sustainable Agriculture', available at http://www.nfu.ca/policy/national-farmers-union-policy-sustainable-agriculture (accessed 5 April 2013).

NFU (2010), 'Farmer's Share of the Retail Food Dollar', available at http://iowafarmersunion.org/wpcontent/uploads/2010/06/june_2010_farmersshare.pdf (accessed 8 October 2012).

Nature's Path (2013), 'Our Roots', available at http://ca-en.naturespath.com/about/our-roots (accessed 17 May 2013).

Nestle, Marion (2004), *Safe Food: Bacteria, Biotechnology, and Bioterrorism*, Berkeley: University of California.

Nestle, Marion (2003), *Food Politics: How the Food Industry Influences Nutrition and Health*, Los Angeles: University of California Press.

Newsweek (1970), 'The Stuff of Life', *Newsweek* May 25, 100.

Nicholson, C. M. Gomez and O. Gao (2011), 'The Costs of Increased Localization for a Multiple-Product Supply Chain: Dairy in the United States', *Food Policy*, **36**, 423–39.

Nielsen, Chantal and Kym Anderson (2000), *Global Market Effects of Alternative European Responses to GMOs*, Copenhagen, CIES Policy Discussion Paper 0032.

Norman, Gurney (1971), 'The Organic Gardening Books', *The (last) Whole Earth Catalog*, 50.

North, Douglass C. (1990), *Institutions, Institutional Change and Economic Performance*, Cambridge: Cambridge University Press.

O'Brien, Robert, Anne Marie Goetz, Jan Aart Scholte and Marc Williams (2000), *Contesting Global Governance: Multilateral Economic Institutions and Global Social Movements*, Cambridge: Cambridge University Press.

Oelhaf, Robert C. (1982), 'Constraints for Commercial Organic Food Production in the USA', in Stuart Hill and Pierre Ott (eds), *Basic Techniques in Ecological Farming: 2nd Annual International Conference held by IFOAM*, Basel: Birkhauser Verlag, pp. 41–2.

Offe, Claus (1985), 'New Social Movements: Changing Boundaries of the Political', *Social Research*, **52**, 817–68.

Ontario Ministry of Agriculture, Food and Rural Affairs (2013), 'Value Chains in Agriculture, Food and Agri-Products Sectors', available at http://www.omafra.gov.on.ca/english/food/valuechains.html (accessed 30 October 2014).

Organic Consumers Association (OCA) (2012), 'About the OCA: Who We Are and What We're Doing', available at http://www.organicconsumers.org/aboutus.cfm (accessed 3 October 2012).

OCA (2007), 'About OCA: Who We Are and What We're Doing', available at http://www.purefood.org/aboutus.htm (accessed 5 August 2012).

OCA (2006a), 'Monsanto Plans to Go Forward with Controversial Terminator Gene Technology', available at https://www.organicconsumers.org/old_articles/monsanto/montreal060222.php (accessed 2 April 2012).

OCA (2006b), 'SOS: Safeguard Organic Standards', available at https://www.organicconsumers.org/old_articles/sos.php (accessed 2 April 2012).

Organic Federation of Canada/Fédération Biologique du Canada (OFC/FBC) (2012), 'Key Goals', available at http://organicfederation.ca/key-goals (accessed 4 October 2012).

Organic Gardening (OG) (1989), 'Sold on Organic', *Organic Gardening*, June, pp. 42–6.

Organic Monitor (2006), 'Hain Celestial Acquires Heinz Subsidiary, *Research News*', available at www.organicmonitor.com/uk.htm#2 (accessed 31 March 2012).

Organic Monitor (2005a), 'Globalisation of the Organic Food Industry, *Research News*', available at www.organicmonitor.com/r2601.htm (accessed 10 September 2010).

Organic Monitor (2005b), 'UK: Whole Foods Market Acquires Fresh & Wild, *Research News*', available at www.organicmonitor.com (accessed 10 September 2005).

Organic Monitor (2003), 'Globalisation of the Organic Food Industry, *Research News*', available at http://www.organicmonitor.com/r1407.htm (accessed 10 September 2012).

Organic Products Regulation (OPR) (2009), 'SOR/2009-176 (2009), Department of Justice Canada', available at http://laws-lois.justice. gc.ca/eng/regulations/SOR-2009-176/FullText.html (accessed 17 September 2012).

Organic Regulatory Committee (ORC) (2003), 'The "Ideal" Canadian Organic Food & Fibre Regulation System: Discussion Paper, Version 1', available at http://organic.usask.ca/ORC%20Updates/Version20I1. pdf (accessed 10 February 2014).

OTA (2014), 'Membership List', available at https://ota.com/member ship/ota-members (accessed 3 October 2014).

OTA (2013), 'OTA Celebrates Successful Annual Fund Campaign', available at https://www.ota.com/news/press-releases/17114 (accessed 15 November 2014).

OTA (2006), 'OTA: Overview', available at https://web.archive.org/web/ 20070814115557/http://www.ota.com/about/accomplishments.html?PHP (accessed 14 September 2014).

OTA (2005), 'Canada, *Research News*', available at https://web.archive. org/web/20060311182535/http://www.ota.com/pp/canada.html?printable =1 (accessed 3 November 2014).

OTA (2004), 'Comparison of EU and US Standards', available at https://web.archive.org/web/20131119203139/http://www.ota.com/ standards/other/eu_us.html (accessed 30 March 2014).

OTA (2002), 'OTA Applauds 2002 Farm Bill for its Boost to Organic Agriculture', available at https://www.organicconsumers.org/old_ articles/Organic/otafarmbill051002.php (accessed 17 September 2012).

Organisation for Economic Co-operation and Development (OECD) (2003a), *The Impact of Regulations on Agri-food Trade: The Technical Barriers to Trade (TBT) and Sanitary and Phytosanitary Measures (SPS) Agreements,* Paris: OECD Publications.

OECD (2003b), *Organic Agriculture: Sustainability, Markets and Policies*, Paris: OECD Publications.

Osteen, Craig et al. (2012), 'Agricultural Resources and Environmental Indicators 2012. Economic Information Bulletin No (EIB-98)', available at http://www.ers.usda.gov/media/874175/eib98.pdf (accessed 9 December 2014).

Padel, Susan (2009), 'The Implementation of Organic Principles and Values in the European Regulation for Organic Food, *Food Policy*', available at http://orgprints.org/5509/1/Padel_et_al_manuscript.pdf (accessed 7 December 2014).

Paige, Jeffrey (1975), *Agrarian Revolutions: Social Movements and Export Agriculture in the Under Developed World*, New York: Free Press.

Parsons, William (2004), 'Organic Fruit and Vegetable Production: Do Farmers Get a Premium Price? *Statistics Canada'*, available at www.statcan/english/freepub/21-004-XIE/21-004-XIE2004103.pdf (accessed 19 February 2004).

Paxton, Angela (1994), *The Food Miles Report: The Dangers of Long-distance Food Transport*, London: SAFE Alliance.

Pedersen, Bjarne (2003), 'Organic Agriculture: The Consumer's Perspective', *OECD: Organic Agriculture: Sustainability, Markets and Policy*, Paris: CABI Publishing, pp. 245–54.

Peters, Suzanne (1979), *The Land in Trust: A Social History of The Organic Farming Movement*, unpublished PhD thesis, Montreal: McGill University.

Petrini, Carlo and Benjamin Watson (eds), (2001), *Slow Food: Collected Thoughts on Taste, Tradition and the Honest Pleasures of Food*, Chelsea Green Publishing Company.

Piven, Frances Fox and Richard Cloward (1977), *Poor People's Movements*, New York: Pantheon.

Pollan, Michael (2009), *In Defense of Food: An Eater's Manifesto*, New York: Penguin Press.

Pollan, Michael (2006), 'Organics Goes Mainstream', *Ideas*, Canadian Broadcasting Corporation (podcasts) 4 September, available at https://web.archive.org/web/20061126111040/http://podcast.cbc.ca/mp3/ideas_20060918_909.mp3 (accessed 2 September 2012).

Pollan, Michael (2001), 'Behind the Organic-Industrial Complex, *New York Times'*, available at http://www.nytimes.com/2001/05/13/magazine/13ORGANIC.html (accessed 8 March 2011).

Powell, Jane (1995), 'Direct Distribution of Organic Produce: Sustainable Food Production in Industrialized Countries', *Outlook on Agriculture*, **24**(2), 3–5.

Public Works and Government Services Canada (PWGSC) (2013), 'Organic Agriculture 32/20', available at http://www.tpsgc-pwgsc.gc.ca/ongc-cgsb/programme-program/normes-standards/comm/32-20-agriculture-eng.html (accessed 14 May 2013).

Qualman, Darrin and Nettie Wiebe (2003), *The Structural Adjustment of Canadian Agriculture*, Ottawa: Canadian Centre for Policy Alternatives.

Raeburn, Paul (1995), *The Last Harvest: The Genetic Gamble that Threatens to Destroy American Agriculture*, Lincoln: University of Nebraska Press.

Ransom, David (2005), 'Fair Trade For Sale', *New Internationalist* April, pp. 34–5.

Raynolds, Laura T. (2004), 'The Globalization of Organic Agri-food Networks', *World Development*, **32**(5), 725–43.

Raynolds, Laura T. (2000), 'Re-embedding Global Agriculture: The International Organic and Fair Trade Movements', *Agriculture and Human Values*, **17**, 297–309.

Redclift, Michael (1997), 'Sustainability and Theory: An Agenda for Action', in David Goodman and Michael Watts (eds), *Globalising Food: Agrarian Questions and Global Restructuring*, New York: Routledge, pp. 333–43.

Reed, Matthew (2010), *Rebels for the Soil: The Rise of the Global Organic Food and Farming Movement*, London: Earthscan.

Reynolds, Cynthia (1999), 'Frankenstein's Harvest,' *Canadian Business* 8 Oct.

Rice, Robert A. (2001), 'Noble Goals and Challenging Terrain: Organic and Fair Trade Coffee Movements in the Global Marketplace', *Journal of Agricultural and Environmental Ethics*, **14**(1), 39–66.

Riddle, James and Lynn Coody (2003), 'Comparison of the EU and US Organic Regulations', in Christina Westermayer and Berwand Greier (eds), *The Organic Guarantee System: The Need and Strategy for Harmonisation and Equivalence*, Bonn: IFOAM, pp. 52–62.

Rigby, D. and D. Caceres (2001), 'Organic Farming and the Sustainability of Agricultural Systems', *Agricultural Systems*, **68**, 21–40.

Roane, Kit R. (2002), 'Ripe for Abuse: Farmworkers say Organic Growers Don't Always Treat Them as Well as They do Your Food, *US New Nation & World* April 22', available at http://ipm.osu.edu/trans/042_221.htm (accessed 3 August 2013).

Rodale Institute (2012), 'Rodale Press and Organic Farming', available at http://www.rodaleinc.com/brand/organic-gardening (accessed 18 November 2012).

Rodale Institute (2006), 'Who We Are', available at https://web.archive.org/web/20070622194516/http://www.rodaleinstitute.org/about/who_body.html (accessed 18 November 2012).

Rodale Press (1971), *Organic Directory*, Emmaus: Rodale Press.

Rodale, J.I. (1959), *Encyclopedia of Organic Gardening*, Emmaus: Rodale Press.

Rodale, J.I. (1959), *How to Grow Vegetables and Fruits by the Organic Method*, Emmaus: Rodale Press.

Rodale, Maria (2010), *Organic Manifesto: How Organic Food Can Heal our Planet, Feed the World and Keep us Safe*, Emmaus: Rodale Press.

Rude, James, Darryl Harrison and Jared Carlberg (2010), 'Market Power in Canadian Beef Packing', *Canadian Journal of Agricultural Economics*, available at http://home.cc.umanitoba.ca/~carlberg/bio/Rude%20Harrison%20Carlberg%20CJAE.pdf (accessed 17 Nov- ember 2013).

Ruiz-Marrero, Carmelo (2004), 'Clouds on the Organic Horizon, *CorpWatch* November 25', available at https://web.archive.org/web/

20090519224348/http://www.globalpolicy.org///socecon/tncs//2004/112
5organic.pdf (accessed 9 May 2012).

Rundgren, Gunnar (2003), 'Introduction', in Christina Westermayer and Berwand Greier (eds), *The Organic Guarantee System: The Need and Strategy for Harmonisation and Equivalence*, Bonn: IFOAM, pp. 6–7.

Ryan, Megan (2001), 'Organic Agriculture: What Does It Have to Offer? *CSIRO Plant Industries*', available at https://web.archive.org/web/20060709201015/http://ifama.org/conferences/2001Conference/Forum Presentations/Ryan_Meg.PDF (accessed 15 May 2013).

Sabatier, Paul A. and Hank C. Jenkins-Smith (1993), 'The Advocacy Coalition Framework: Assessment, Revisions and Implications for Scholars and Practitioners', *Policy Change and Learning: An Advocacy Coalition Approach*, Boulder: Westview Press.

Schaeffer, Robert (1995), 'Free Trade Agreements: Their Impact on Agriculture and the Environment', in Philip McMichael (ed.), *Food and Agrarian Orders in the World Economy*, Westport: Greenwood Press, pp. 255–76.

Schlosser, Eric (2002), *Fast Food Nation: The Dark Side of the All-American Meal*, New York: Perennial Press.

Schmeiser, Percy (2007), 'Monsanto vs. Schmeiser: The Classic David and Goliath Struggle', available at http://www.mindfully.org/GE/2003/Monsanto-vs-Schmeiser8may03.htm (accessed 14 June 2010).

Schneider, Keith (1989), 'Big Farm Companies Try Hand at Organic Methods', *New York Times* 28 May 1989, Business Insights, (accessed 28 August 2013).

Schumacher, E.F. (1973), *Small is Beautiful*, London: Blond and Briggs.

Scialabba, Nadia (1999), 'Special: Organic Agriculture and the FAO, SD dimensions, October 1999', available at https://web.archive.org/web/20130329160754/http://www.fao.org/sd/epdirect/epre0055.htm (accessed 23 October 2012).

Seiff, Joanne (2005), 'Why Bother Certifying Organic? The Argument Against USDA Certification', *Organic Producer Magazine* Sept/Oct, https://web.archive.org/web/20060623072320/http://www.organicproducer mag. com/getMoreInfo.cfm?SID=824&level=1 (accessed 11 October 2010).

Severson, Kim and Andrew Martin (2009), 'It's Organic, but Does That Mean It's Safer?', *New York Times* (Dining and Wine), 3 March, available at http://www.nytimes.com/2009/03/04/dining/04cert.html?pagewanted=all&_r=0 (accessed 3 October 2012).

Shapouri, Hosein, James Duffield, Andrew McAloon and Michael Wang (2004), *The 2001 Net Energy Balance of Corn-Ethanol*, U.S. Department of Agriculture, Office of Chief Economist (OCE), Agricultural Research Service (ARS), Washington DC', available at https://web.

archive.org/web/20121010040134/http://www.usda.gov/oce/reports/energy/
net_energy_balance.pdf (accessed 8 November 2013).
Sharpin, Steven (2006), 'Paradise Sold: What are You Buying When You
Buy Organic?', *The New Yorker*, 5 August.
Shreck, Aimee, et al. (2006), 'Social Sustainability, Farm Labor, and
Organic Agriculture: Findings from an Exploratory Analysis', *Agriculture and Human Values*, **23**(4), 439–56.
Shiva, Vandana (2000), *Stolen Harvest: The Highjacking of the Global
Food Supply*, Boston: South End Press.
Sikkink, Kathryn (2005), 'Patterns of Dynamic Multilevel Governance
and the Insider-Outsider Coalition', in Donatella della Porta and Sidney
Tarrow (eds), *Transnational Protest and Global Activism*, Toronto:
Rowman & Littlefield Publishers.
Singer, Peter and Jim Mason (2006), *The Way We Eat: Why Our Food
Choices Matter*, Emmaus: Rodale Press.
Skogstad, Grace (1987), *The Politics of Agricultural Policy-Making in
Canada*, Toronto: University of Toronto Press.
Sligh, Michael and T. Cierpka (2007), 'Organic Values', in William
Lockeretz (ed.), *Organic Farming: An International History*, Wallingford: CABI, pp. 30–39.
Sligh, Michael and Carolyn Christman (2003), 'Who Owns Organic? The
Global Status, Prospects and Challenges of a Changing Organic
Market', *Rural Advancement Foundation International USA-2003*,
available at http://rafiusa.org/blog/who-owns-organic-the-global-status-prospects-and-challenges-of-a-changing-organic-market/ (accessed 10
February 2014).
Smith, Adrian et al. (2002), 'Networks of Value, Commodities and
Regions: Reworking Divisions of Labour in Macro-regional Economies', *Progress in Human Geography*, **26**(1), 41–63.
Smith, Alisa and J.B. MacKinnon (2007), *The 100 Mile Diet: A Year of
Local Eating*, Toronto: Random House.
Smith, Alison et al. (2005), *The Validity of Food Miles as an Indicator of
Sustainable Development*, final report, London: DEFRA.
Snow, David A. et al. (1986), 'Frame Alignment Processes, Micro-mobilization, and Movement Participation', *American Sociological
Review*, **51**, 464–81.
The Soil Association (2013), 'Our History', available at http://www.soil
association.org/aboutus/ourhistory (accessed 27 August 2013).
The Soil Association and Sustain (2001), 'Organic Food and Farming:
Myth and Reality, Soil Association', available at http://www.soil
association.org/LinkClick.aspx?fileticket=30Bk3Sg6Pp0%3D&tabid=
385 (accessed 24 February 2013).

232   *The changing politics of organic food in North America*

Sparling, David and Roberta Cook (2000), 'Strategic Alliances and Joint Ventures Under NAFTA: Concepts and Evidence', *Policy Harmonization and Adjustment in the North American Agricultural and Food Industry*, proceedings of the Fifth Agricultural and Food policy Systems Workshop, pp. 68–94, available at https://web.archive.org/web/20051226230424/http://www.farmfoundation.org/maroon/sparling.pdf (accessed 19 November 2011).

Stanton, Gretchen (2004), 'A Review of the Operation of the Agreement on Sanitary and Phyto-Sanitary Measures', in Merlina D. Ingco and L. Alan Winters (eds), *Agriculture and the New Trade Agenda: Creating a Global Trading Environment for Development*, Cambridge: Cambridge University Press, pp. 101–10.

Statistics Canada (2011), '2011 Census of Agriculture', available at http://www.statcan.gc.ca/daily-quotidien/120510/dq120510a-eng.htm (accessed 2 June 2012).

Statistics Canada (2007), 'Snapshot of Canadian Agriculture, Census of Agriculture, 2006', available at http://www.statcan.gc.ca/ca-ra2006/articles/snapshot-portrait-eng.htm (accessed 7 June 2011).

Statistics Canada (2006), 'Summary Table for the Census of Agriculture', available at http://www.statcan.gc.ca/tables-tableaux/sum-som/l01/ind 01/l3_920-eng.htm?hili_none (accessed 7 May 2009).

Statistics Canada (2005), 'Farm Holdings, Census Data, Canada by Province, 1871–1971, Statistics Canada Series M12-22 (Archived)', available at www.statcan.ca/english/freepub/11-516-XIE/sectopmm/M12-22.csv (accessed 2 September 2010).

Statistics Canada (2004), 'Census of Agriculture: Historical Perspective, Statistics Canada', available at http://www.statcan.gc.ca/pub/95-632-x/2007000/4129762-eng.htm (accessed 10 September 2013).

Statistics Canada (2002a), 'Census of Agriculture 2002', Ottawa: Government of Canada, available at https://web.archive.org/web/20040604194220/http://www.statcan.ca/english/Pgdb/econ103a.htm (accessed 12 February 2012).

Statistics Canada (2002b), 'Farmers Leaving the Field', *Perspectives,* February, Ottawa: Government of Canada.

Steele, J. (1995), *Local Food Links: New Ways of Getting Organic Food from Farm to Table*, Bristol: Soil Association.

Steffen, Robert et al. (eds), (1972), *Organic Farming: Methods and Markets: An Introduction to Ecological Agriculture*, Emmaus, PA: Rodale Press.

Steiner, R. (1924), 'Report to Members of the Anthroposophical Society after the Agriculture Course', Dornach, Switzerland, 20 June, translated by C.E. Creeger and M. Gardner, in M. Gardner, *Spiritual*

Foundations for the Renewal of Agriculture by Rudolf Steiner, (1993) Kimberton: Bio-Dynamic Farming and Gardening Association, pp. 1–12.

Stephenson, Lorraine (2007), 'Organic Imports Satisfy Consumer Demand', *Manitoba Co-operator*, 15 March.

Stoker G. (1998), 'Governance as a Theory: Five Propositions', *International Social Science Journal*, **155**, 17–28.

Stringer, Christina (2006), 'Forest Certification and Global Commodity Chains', *Journal of Economic Geography*, **6**, 701–22.

Sumner, Jennifer (2005), 'Organic Intellectuals: Lifelong Learning in the Organic Farm Movement', *Canadian Association for the Study of Adult Education,* Conference Proceedings 28–31 May, available at https://web.archive.org/web/20090830125626/http://www.oise.utoronto.ca/CASAE/cnf2005/2005onlineProceedings/CAS2005Pro-Sumner.pdf (accessed 24 July 2012).

SunOpta (2004), *Gathering Momentum: 2003 Annual Report*, March.

Swanton Berry Farms, (2012), 'The Farm', available at http://www.swantonberryfarm.com/pages/farm_general.html (accessed 1 November 2012).

Synovate and Julie Winam (2003), *COABC Market Survey for Certified Organic*, Vancouver: Associations of British Columbia.

Tarrow, Sidney (1983), *Struggling to Reform: Social Movements and Policy Change During Cycles of Protest*, New York: Cornell University Press.

Tarrow, Sidney (2005), *The New Transnational Activism*, New York: Cambridge University Press.

Tate, William B. (1994), 'The Development of the Organic Industry and Market: An International Perspective', *The Economics of Organic Farming: An International Perspective*, Wallingford: CAB International, pp. 11–25.

Thomson, Gary D. (1998), 'Consumer Demand for Organic Foods: What We Know and What We Need To Know', *American Journal of Agricultural Economics*, **80**(5), 1113–18.

Tick, Paul (2004), 'Big Business Enters the Organic Market Place', *Organic Consumers Association*', available at www.organicconsumers.org/organic/big_business.cfm (accessed 12 February 2009).

Tickner, J. Ann (1993), 'States and Markets: An Ecofeminist Perspective on IPE', *International Political Science Review*, **14**(1), 39–69.

Tilly, Charles (1978), *From Mobilization to Revolution*, Reading: Addison-Wesley.

Tilly, Charles (2004), *Social Movements, 1768-2004*, Boulder: Paradigm Publishers.

Trujillo, Elizabeth (2012), 'The WTO Appellate Body Knocks Down U.S. "Dolphin-Safe" Tuna Labels But Leaves a Crack for PPMs', *ASIL Insights*, **16**(25), available at http://www.asil.org/sites/default/files/insight120726.pdf (accessed 5 May 2013).

Tuomisto, H.L., I.D. Hodge, P. Riordan and D.W. Macdonald (2012), 'Does Organic Farming Reduce Environmental Impacts? – A Meta-analysis of European Research', *Journal of Environmental Management*, **112**, 309–20.

United Nations Committee on Trade and Development (UNCTAD), FAO and IFOAM (2009), 'Harmonization and Equivalence in Organic Agriculture. Vol. 6', available at http://www.ifoam.org/partners/projects/pdfs/ITFVol6.pdf (accessed 14 September 2012).

United States Census Bureau (USCB) (2012), 'Farms: Numbers and Acreage', available at https://www.census.gov/compendia/statab/2012/tables/12s0824.pdf (accessed 12 October 2012).

United States Congress, Office of Technology Assessment (2005), *Agriculture, Trade, and Environment: Achieving Complementary Policies*, OTA-ENV-617. Washington, DC: US Government Printing Office.

United States Department of Agriculture (USDA) (2014), '2014 Farm Bill Highlights', available at http://www.usda.gov/documents/usda-2014-farm-bill-highlights.pdf (accessed 12 January 2015).

USDA/Economic Research Service (USDA/ERS) (2010), 'Organic Production', available at http://www.ers.usda.gov/data-products.aspx (accessed 22 September 2012).

USDA (2006), 'Table 2: U.S. Certified Organic Farm Land Acreage, Livestock Numbers, and Farm Operations, 1992–2005, ERS/USDA', available at www.ers/usda.gov/Data/Organic/Data/Farmland%20livestock%20and%20farm%20ops%2092-05 (accessed 7 September 2013).

USDA (2005), 'U.S. Market Profile for Organic Food Products, *Foreign Agriculture Service*', available at http://www.fas.usda.gov/agx/organics/USMarketProfileOrganicFoodFeb2005.pdf (accessed 5 November 2013).

USDA (2004), 'Number of Farms, Average Size of Farm, and Land in Farms in the United States, 1974–2003', available at http://www.nass.usda.gov/Ky/B2004/p010.pdf (accessed 2 September 2013).

USDA (2002a), 'Canada Organic Foods Industry Report', *Foreign Agricultural Service. GAIN Report # CA2001*.

USDA (2002b), 'Historical Highlights: 2002 and Earlier Census Years', *2002 Census of Agriculture – United States Data*', USDA, National Agricultural Statistics Service.

USDA (2002c), 'National Organic Program: Applicability: Preamble, *AMS/USDA*', available at http://www.ams.usda.gov/nop/NOP/standards/ApplicPre.html (accessed 4 October 2013).

USDA (2002d), 'National Organic Program, *AMS/USDA*', available at http://www.ams.usda.gov/nop/NOP/standards/FullText.pdf (accessed 10 January 2012).

USDA (1999), 'ISO 65 Accreditation For Organic Certification Bodies, *MGC Instruction 707*. 23 August', available at https://web.archive.org/web/20061003024239/http://www.ams.usda.gov//lsg/arc//707.pdf (accessed 21 July 2012).

USDA (1990), 'The Organic Foods Production Act of 1990, *National Organic Program, Agricultural Marketing Service*', available at https://web.archive.org/web/20111102181654/http://agriculture.senate.gov/Legislation/Compilations/AgMisc/OGFP90.pdf (accessed 3 April 2011).

USDA/ERS (2003), 'USDA Accredited Organic Certification Programs Active in 2002 and 2003', available at http://www.ers.usda.gov/Data/Organic/data/certifiers.xls (accessed 5 November 2013).

Urwin, D. (1986), 'Responsibility More Than Chemical Free', *New Farmer and Grower*, 12 pp. 10–11.

van Elzakker, Bo (2003), 'International Harmonisation and Equivalence in Organic Agriculture: What IFOAM and the IOAS can Contribute', in Christina Westermayer and Berwand Greier (eds), *The Organic Guarantee System: The Need and Strategy for Harmonisation and Equivalence*, Bonn: IFOAM, pp. 82–4.

Van Praet (2013), 'Whole Foods co-CEO says Grocer Wants to Open 40 More Stores in Canada' *Financial Post*, May 22, 2013', available at http://business.financialpost.com/2013/05/22/whole-foods-says-grocer-wants-to-open-40-more-canadian-stores/ (accessed 12 January 2015).

Vaupel, Suzanne and Gunnar Rundgren (2003), 'The Interface Between the IFOAM and International Organic Guarantee System and Regulations', in Christina Westermayer and Berwand Greier (eds), *The Organic Guarantee System: The Need and Strategy for Harmonisation and Equivalence*, Bonn: IFOAM, pp. 96–9.

Vos, Timothy (2000), 'Visions of the Middle Landscape: Organic Farming and the Politics of Nature', *Agriculture and Human Values*, **17**, 245–56.

Vossenaar, Rene (2003), 'Promoting Production and Exports of Organic Agriculture in Developing Countries', in Christina Westermayer and Berwand Greier (eds), *The Organic Guarantee System: The Need and Strategy for Harmonisation and Equivalence*, Bonn: IFOAM, pp. 10–15.

Walnut Acres (2005), 'Certified Organic Future: National Consumer Survey', available at www.walnutacres.com (accessed 23 February 2010).

Washington Post (1974), 'Organic Farming "Scientific Nonsense"', *Washington Post*, 28 February.

Warnock, John W. (1987), *The Politics of Hunger: The Global Food System*, Toronto: Methuen Press.

Weeks, Carly (2006), 'Organics Industry goes Mainstream', *The Vancouver Sun*, April 17, D4.

Welling, Andrea (1999), *Feeding Our Communities: A Feminist Perspective on the Challenge of Organic Food Production for Women in B.C.*, unpublished MA thesis, Burnaby: Simon Fraser University.

Westermayer, Christina and Berwand Greier (eds) (2003), *The Organic Guarantee System: The Need and Strategy for Harmonisation and Equivalence*, Bonn: IFOAM.

Whatmore, Sarah (2002), 'From Farming to Agribusiness: Global Agrifood Networks', in R.J. Johnston et al. (eds.), *Geographies of Global Change: Remapping the World*, New York: Blackwell Publishing, pp. 57–67.

White, Hilda (1972), 'The Organic Foods Movement: What It Is, and What the Food Industry Should Do about It', *Food Technology*, April, 29–33.

Willer, Helga and Julia Lernoud (eds) (2014), *The World of Organic Agriculture – Statistics and Emerging Trends 2014*, Research Institute of Organic Agriculture (FiBL), Frick, and IFOAM, Bonn.

Willer, Helga and Julia Lernoud (2013), 'Current Statistics on Organic Agriculture Worldwide: Organic Area, Producers and Market', in Helga Willer and Julia Lernoud (eds), *The World of Organic Agriculture: Statistics and Emerging Trends 2013*, Research Institute of Organic Agriculture (FiBL), Frick and IFOAM, Bonn: 36–108.

Willer, Helga and Lukas Kilcher (eds) (2009), *The World of Organic Agriculture – Statistics and Emerging Trends 2009*, Research Institute of Organic Agriculture (FiBL), Frick and IFOAM, Bonn.

Willer, Helga and Minou Yuseffi (eds) (2005), *The World of Organic Agriculture*, Research Institute of Organic Agriculture (FiBL), Frick, and IFOAM, Bonn.

Willer, Helga and Minou Yuseffi (eds) (2004), *The World of Organic Agriculture,* Research Institute of Organic Agriculture (FiBL), Frick, and IFOAM, Bonn.

Williams, Mark (2005), 'Civil Society and the World Trading System', in Dominic Kelly and Wynn Grant (eds), *The Politics of International Trade in the Twenty-First Century: Actors, Issues and Regional Dynamics*, Toronto: Palgrave-MacMillan, pp. 30–46.

The Whole Earth Catalog (WEC) (1970), 'Function and Purpose', *The Whole Earth Catalog* [cover] Sept.

WEC (1970), 'Mother Earth News', *The Whole Earth Catalog*, Jan, 19.

Whole Earth Magazine (2007), 'About Whole Earth', available at https://web.archive.org/web/20070816103657/http://www.wholeearthmag.com/about.html (accessed 15 January 2007).

Whole Foods Markets (WFM) (2015), 'Newsroom Home', available at http://media.wholefoodsmarket.com/faq/ (accessed 12 January 2015).

WFM (2013), Annual Report, available at http://www.wholefoodsmarket.com/sites/default/files/media/Global/Company%20Info/PDFs/WFM-2013-Annual-Stakeholders-Report.pdf (accessed 19 March 2015).

WFM (2005a), 'Whole Foods Market-Company Timeline', available at www.wholefoodsmarket.com/company/timeline.html (accessed 10 September 2011).

WFM (2005b), 'Annual Report', available at http://www.wholefoodsmarket.com/investor/ar05.pdf (accessed 29 April 2010).

Whole Workers Unite! (2006), 'Whole Foods Workers Unite!' available at https://wcb.archive.org/web/20070810210900/http://www.wholeworkersunite.org/node (accessed 29 April 2012).

Wolfe, Robert (1998), *Farm Wars: The Political Economy of Agriculture and the International Trade Regime*, New York: St. Marten's Press.

Wolnak, B. (1972), 'Health Foods: Natural Basic and Organic', *Food Drug Cosmetics Law Journal*, **27**, 452–60.

World Health Organization (WHO) (2007), 'Food Safety and Food Borne Illness, Fact Sheet no. 237', available at https://web.archive.org/web/20090215152021/http://www.who.int/mediacentre/factsheets/fs237/en/print.html (accessed 2 July 2010).

WTO (2014a), 'Dispute Settlement: Dispute DS26, European Communities – Measures Concerning Meat and Meat Products (Hormones)', available at http://www.wto.org/english/tratop_e/dispu_e/cases_e/ds26_e.htm (accessed 5 December 2014).

WTO (2014b), 'Agreement on Implementation of Article VI of the General Agreement on Tariffs and Trade, 1994', available at http://www.wto.org/english/docs_e/legal_e/19-adp_01_e.htm (accessed 4 September 2014).

WTO (2013a), 'Legal Texts: The WTO agreements: A Summary of the Final Act of the Uruguay Round', available at www.wto.org/english/docs_e/legal_e/ursum_e.htm#dAgreement (accessed 7 November 2013).

WTO (2013b), 'United States – Subsidies on Upland Cotton, (Dispute Settlement DS267)', available at http://www.wto.org/English/tratop_e/dispu_e/cases_e/ds267_e.htm (accessed 19 July 2013).

WTO (2012a), 'Agreement on Technical Barriers To Trade (TBT)', available at http://www.wto.org/english/docs_e/legal_e/17-tbt.pdf (accessed 20 July 2012).

WTO (2012b), 'Sanitary and Phyto-Sanitary Measures', available at www.wto.org/english/tratop-e/sps/spe-e.htm (accessed 14 June 2012).

WTO (2012c), 'Technical Explanation, Technical Information on the Technical Barriers To Trade, *Agreement on Technical Barriers To Trade*', available at http://www.wto.org/english/tratop_e/tbt_e/tbt_info_e.htm (accessed 20 July 2012).

WTO (2012d), 'Environment: Issues; Environmental Requirements and Market Access: Preventing "Green Protectionism"', available at http://www.wto.org/english/tratop_e/envir_e/envir_req_e.htm (accessed 20 October 2012).

WTO (2012e), 'Environmental Issues: Labelling', available at http://www.wto.org/english/tratop_e/envir_e/labelling_e.htm (accessed 3 November 2012).

WTO (2011), 'Environment: Dispute 4, Mexico versus US: "Tuna–Dolphin, 1992"', available at http://www.wto.org/English/tratop_e/envir_e/edis04_e.htm (accessed 20 July 2011).

Wood, Ellen Meiksins (1999), *The Empire of Capital*, London: Verson.

Wunsch, Patti (2003), *There's More to Organic Farming Than Pesticide-Free*, Statistics Canada, Catalogue no. 96 – 325-XPB.

Youngberg, Garth and Fredrick H. Buttel (1984), 'Public Policy and Socio-Political Factors Affecting the Future of Sustainable Farming Systems', *Organic Farming: Current Technology and Its Role in a Sustainable Agriculture*, ASA special publication no. 46. Madison: American Society of Agronomy, pp. 167–85.

Index